Sheldon Axler

KU-017-535

# Linear Algebra Done Right

## Second Edition

Springer

Sheldon Axler
Mathematics Department
San Francisco State University
San Francisco, CA 94132
USA

*Editorial Board*

S. Axler
Mathematics Department
San Francisco
    State University
San Francisco, CA 94132
USA

F.W. Gehring
Department of
    Mathematics
University of Michigan
Ann Arbor, MI 48109
USA

P.R. Halmos
Department of
    Mathematics
Santa Clara University
Santa Clara, CA 95053
USA

Mathematics Subject Classification (1991): 15-01

Library of Congress Cataloging-in-Publication Data
Axler, Sheldon Jay
    Linear algebra done right / Sheldon Axler. – 2nd ed.
        p.  cm. — (Undergraduate texts in mathematics)
    Includes index.
    ISBN 0-387-98259-0 (alk. paper). – ISBN 0-387-98258-2 (pbk. :
alk. paper)
    1. Algebra, Linear.  I. Title.  II. Series.
QA184.A96    1997
512'.5—dc20                                            97-16664

Printed on acid-free paper.

© 1997, 1996 Springer-Verlag New York, Inc.
All rights reserved. This work may not be translated or copied in whole or in part
without the written permission of the publisher (Springer-Verlag New York, Inc., 175
Fifth Avenue, New York, NY 10010, USA), except for brief excerpts in connection with
reviews or scholarly analysis. Use in connection with any form of information storage
and retrieval, electronic adaptation, computer software, or by similar or dissimilar
methodology now known or hereafter developed is forbidden.
The use of general descriptive names, trade names, trademarks, etc., in this publica-
tion, even if the former are not especially identified, is not to be taken as a sign that
such names, as understood by the Trade Marks and Merchandise Marks Act, may
accordingly be used freely by anyone.

Production managed by Victoria Evaretta; manufacturing supervised by Jeffrey Taub.
Photocomposed copy prepared from the author's TeX files.
Printed and bound by Hamilton Printing Co., Rensselaer, NY.
Printed in the United States of America.

9 8 7 6 5 4 3 2 1

ISBN 0-387-98259-0  Springer-Verlag New York Berlin Heidelberg  SPIN 10629393  (hardcover)
ISBN 0-387-98258-2  Springer-Verlag New York Berlin Heidelberg  SPIN 10629369  (softcover)

# Undergraduate Texts in Mathematics

*Editors*

S. Axler

F.W. Gehring

P. Halmos

2

**Springer**

*New York*
*Berlin*
*Heidelberg*
*Barcelona*
*Budapest*
*Hong Kong*
*London*
*Milan*
*Paris*
*Santa Clara*
*Singapore*
*Tokyo*

LIVERPOOL JMU LIBRARY

3  1111  00599  8677

LIVERPOOL
JOHN MOORES UNIVERSITY
AVRIL ROBARTS LRC
TITHEBARN STREET
LIVERPOOL L2 2ER
TEL. 0151 231 4022

# Undergraduate Texts in Mathematics

*(continued after index)*

# Contents

# *Preface to the Instructor*

You are probably about to teach a course that will give students their second exposure to linear algebra. During their first brush with the subject, your students probably worked with Euclidean spaces and matrices. In contrast, this course will emphasize abstract vector spaces and linear maps.

The audacious title of this book deserves an explanation. Almost all linear algebra books use determinants to prove that every linear operator on a finite-dimensional complex vector space has an eigenvalue. Determinants are difficult, nonintuitive, and often defined without motivation. To prove the theorem about existence of eigenvalues on complex vector spaces, most books must define determinants, prove that a linear map is not invertible if and only if its determinant equals 0, and then define the characteristic polynomial. This tortuous (torturous?) path gives students little feeling for why eigenvalues must exist.

In contrast, the simple determinant-free proofs presented here offer more insight. Once determinants have been banished to the end of the book, a new route opens to the main goal of linear algebra—understanding the structure of linear operators.

This book starts at the beginning of the subject, with no prerequisites other than the usual demand for suitable mathematical maturity. Even if your students have already seen some of the material in the first few chapters, they may be unaccustomed to working exercises of the type presented here, most of which require an understanding of proofs.

- Vector spaces are defined in Chapter 1, and their basic properties are developed.

- Linear independence, span, basis, and dimension are defined in Chapter 2, which presents the basic theory of finite-dimensional vector spaces.

- Linear maps are introduced in Chapter 3. The key result here is that for a linear map $T$, the dimension of the null space of $T$ plus the dimension of the range of $T$ equals the dimension of the domain of $T$.

- The part of the theory of polynomials that will be needed to understand linear operators is presented in Chapter 4. If you take class time going through the proofs in this chapter (which contains no linear algebra), then you probably will not have time to cover some important aspects of linear algebra. Your students will already be familiar with the theorems about polynomials in this chapter, so you can ask them to read the statements of the results but not the proofs. The curious students will read some of the proofs anyway, which is why they are included in the text.

- The idea of studying a linear operator by restricting it to small subspaces leads in Chapter 5 to eigenvectors. The highlight of the chapter is a simple proof that on complex vector spaces, eigenvalues always exist. This result is then used to show that each linear operator on a complex vector space has an upper-triangular matrix with respect to some basis. Similar techniques are used to show that every linear operator on a real vector space has an invariant subspace of dimension 1 or 2. This result is used to prove that every linear operator on an odd-dimensional real vector space has an eigenvalue. All this is done without defining determinants or characteristic polynomials!

- Inner-product spaces are defined in Chapter 6, and their basic properties are developed along with standard tools such as orthonormal bases, the Gram-Schmidt procedure, and adjoints. This chapter also shows how orthogonal projections can be used to solve certain minimization problems.

- The spectral theorem, which characterizes the linear operators for which there exists an orthonormal basis consisting of eigenvectors, is the highlight of Chapter 7. The work in earlier chapters pays off here with especially simple proofs. This chapter also deals with positive operators, linear isometries, the polar decomposition, and the singular-value decomposition.

- The minimal polynomial, characteristic polynomial, and generalized eigenvectors are introduced in Chapter 8. The main achievement of this chapter is the description of a linear operator on a complex vector space in terms of its generalized eigenvectors. This description enables one to prove almost all the results usually proved using Jordan form. For example, these tools are used to prove that every invertible linear operator on a complex vector space has a square root. The chapter concludes with a proof that every linear operator on a complex vector space can be put into Jordan form.

- Linear operators on real vector spaces occupy center stage in Chapter 9. Here two-dimensional invariant subspaces make up for the possible lack of eigenvalues, leading to results analogous to those obtained on complex vector spaces.

- The trace and determinant are defined in Chapter 10 in terms of the characteristic polynomial (defined earlier without determinants). On complex vector spaces, these definitions can be restated: the trace is the sum of the eigenvalues and the determinant is the product of the eigenvalues (both counting multiplicity). These easy-to-remember definitions would not be possible with the traditional approach to eigenvalues because that method uses determinants to prove that eigenvalues exist. The standard theorems about determinants now become much clearer. The polar decomposition and the characterization of self-adjoint operators are used to derive the change of variables formula for multivariable integrals in a fashion that makes the appearance of the determinant there seem natural.

This book usually develops linear algebra simultaneously for real and complex vector spaces by letting **F** denote either the real or the complex numbers. Abstract fields could be used instead, but to do so would introduce extra abstraction without leading to any new linear algebra. Another reason for restricting attention to the real and complex numbers is that polynomials can then be thought of as genuine functions instead of the more formal objects needed for polynomials with coefficients in finite fields. Finally, even if the beginning part of the theory were developed with arbitrary fields, inner-product spaces would push consideration back to just real and complex vector spaces.

Even in a book as short as this one, you cannot expect to cover everything. Going through the first eight chapters is an ambitious goal for a one-semester course. If you must reach Chapter 10, then I suggest covering Chapters 1, 2, and 4 quickly (students may have seen this material in earlier courses) and skipping Chapter 9 (in which case you should discuss trace and determinants only on complex vector spaces).

A goal more important than teaching any particular set of theorems is to develop in students the ability to understand and manipulate the objects of linear algebra. Mathematics can be learned only by doing; fortunately, linear algebra has many good homework problems. When teaching this course, I usually assign two or three of the exercises each class, due the next class. Going over the homework might take up a third or even half of a typical class.

A solutions manual for all the exercises is available (without charge) only to instructors who are using this book as a textbook. To obtain the solutions manual, instructors should send an e-mail request to me (or contact Springer if I am no longer around).

Please check my web site for a list of errata (which I hope will be empty or almost empty) and other information about this book.

I would greatly appreciate hearing about any errors in this book, even minor ones. I welcome your suggestions for improvements, even tiny ones. Please feel free to contact me.

Have fun!

Sheldon Axler
Mathematics Department
San Francisco State University
San Francisco, CA 94132, USA

e-mail: axler@math.sfsu.edu
www home page: http://math.sfsu.edu/axler

# *Preface to the Student*

You are probably about to begin your second exposure to linear algebra. Unlike your first brush with the subject, which probably emphasized Euclidean spaces and matrices, we will focus on abstract vector spaces and linear maps. These terms will be defined later, so don't worry if you don't know what they mean. This book starts from the beginning of the subject, assuming no knowledge of linear algebra. The key point is that you are about to immerse yourself in serious mathematics, with an emphasis on your attaining a deep understanding of the definitions, theorems, and proofs.

You cannot expect to read mathematics the way you read a novel. If you zip through a page in less than an hour, you are probably going too fast. When you encounter the phrase "as you should verify", you should indeed do the verification, which will usually require some writing on your part. When steps are left out, you need to supply the missing pieces. You should ponder and internalize each definition. For each theorem, you should seek examples to show why each hypothesis is necessary.

Please check my web site for a list of errata (which I hope will be empty or almost empty) and other information about this book.

I would greatly appreciate hearing about any errors in this book, even minor ones. I welcome your suggestions for improvements, even tiny ones.

Have fun!

Sheldon Axler
Mathematics Department
San Francisco State University
San Francisco, CA 94132, USA

e-mail: axler@math.sfsu.edu
www home page: http://math.sfsu.edu/axler

# *Acknowledgments*

I owe a huge intellectual debt to the many mathematicians who created linear algebra during the last two centuries. In writing this book I tried to think about the best way to present linear algebra and to prove its theorems, without regard to the standard methods and proofs used in most textbooks. Thus I did not consult other books while writing this one, though the memory of many books I had studied in the past surely influenced me. Most of the results in this book belong to the common heritage of mathematics. A special case of a theorem may first have been proved in antiquity (which for linear algebra means the nineteenth century), then slowly sharpened and improved over decades by many mathematicians. Bestowing proper credit on all the contributors would be a difficult task that I have not undertaken. In no case should the reader assume that any theorem presented here represents my original contribution.

Many people helped make this a better book. For useful suggestions and corrections, I am grateful to William Arveson (for suggesting the proof of 5.13), Marilyn Brouwer, William Brown, Robert Burckel, Paul Cohn, James Dudziak, David Feldman (for suggesting the proof of 8.40), Pamela Gorkin, Aram Harrow, Pan Fong Ho, Dan Kalman, Robert Kantrowitz, Ramana Kappagantu, Mizan Khan, Mikael Lindström, Jacob Plotkin, Elena Poletaeva, Mihaela Poplicher, Richard Potter, Wade Ramey, Marian Robbins, Jonathan Rosenberg, Joan Stamm, Thomas Starbird, Jay Valanju, and Thomas von Foerster.

Finally, I thank Springer for providing me with help when I needed it and for allowing me the freedom to make the final decisions about the content and appearance of this book.

# CHAPTER 1

# *Vector Spaces*

Linear algebra is the study of linear maps on finite-dimensional vector spaces. Eventually we will learn what all these terms mean. In this chapter we will define vector spaces and discuss their elementary properties.

In some areas of mathematics, including linear algebra, better theorems and more insight emerge if complex numbers are investigated along with real numbers. Thus we begin by introducing the complex numbers and their basic properties.

# *Complex Numbers*

*The symbol i was first used to denote $\sqrt{-1}$ by the Swiss mathematician Leonhard Euler in 1777.*

You should already be familiar with the basic properties of the set $\mathbf{R}$ of real numbers. Complex numbers were invented so that we can take square roots of negative numbers. The key idea is to assume we have a square root of $-1$, denoted $i$, and manipulate it using the usual rules of arithmetic. Formally, a ***complex number*** is an ordered pair $(a, b)$, where $a, b \in \mathbf{R}$, but we will write this as $a + bi$. The set of all complex numbers is denoted by $\mathbf{C}$:

$$\mathbf{C} = \{a + bi : a, b \in \mathbf{R}\}.$$

If $a \in \mathbf{R}$, we identify $a + 0i$ with the real number $a$. Thus we can think of $\mathbf{R}$ as a subset of $\mathbf{C}$.

Addition and multiplication on $\mathbf{C}$ are defined by

$$(a + bi) + (c + di) = (a + c) + (b + d)i,$$
$$(a + bi)(c + di) = (ac - bd) + (ad + bc)i;$$

here $a, b, c, d \in \mathbf{R}$. Using multiplication as defined above, you should verify that $i^2 = -1$. Do not memorize the formula for the product of two complex numbers; you can always rederive it by recalling that $i^2 = -1$ and then using the usual rules of arithmetic.

You should verify, using the familiar properties of the real numbers, that addition and multiplication on $\mathbf{C}$ satisfy the following properties:

**commutativity**
  $w + z = z + w$ and $wz = zw$ for all $w, z \in \mathbf{C}$;

**associativity**
  $(z_1 + z_2) + z_3 = z_1 + (z_2 + z_3)$ and $(z_1 z_2)z_3 = z_1(z_2 z_3)$ for all $z_1, z_2, z_3 \in \mathbf{C}$;

**identities**
  $z + 0 = z$ and $z1 = z$ for all $z \in \mathbf{C}$;

**additive inverse**
  for every $z \in \mathbf{C}$, there exists a unique $w \in \mathbf{C}$ such that $z + w = 0$;

**multiplicative inverse**
  for every $z \in \mathbf{C}$ with $z \neq 0$, there exists a unique $w \in \mathbf{C}$ such that $zw = 1$;

**distributive property**

$\lambda(w + z) = \lambda w + \lambda z$ for all $\lambda, w, z \in \mathbf{C}$.

For $z \in \mathbf{C}$, we let $-z$ denote the additive inverse of $z$. Thus $-z$ is the unique complex number such that

$$z + (-z) = 0.$$

Subtraction on $\mathbf{C}$ is defined by

$$w - z = w + (-z)$$

for $w, z \in \mathbf{C}$.

For $z \in \mathbf{C}$ with $z \neq 0$, we let $1/z$ denote the multiplicative inverse of $z$. Thus $1/z$ is the unique complex number such that

$$z(1/z) = 1.$$

Division on $\mathbf{C}$ is defined by

$$w/z = w(1/z)$$

for $w, z \in \mathbf{C}$ with $z \neq 0$.

So that we can conveniently make definitions and prove theorems that apply to both real and complex numbers, we adopt the following notation:

> Throughout this book,
> **F** stands for either **R** or **C**.

Thus if we prove a theorem involving **F**, we will know that it holds when **F** is replaced with **R** and when **F** is replaced with **C**. Elements of **F** are called *scalars*. The word "scalar", which means number, is often used when we want to emphasize that an object is a number, as opposed to a vector (vectors will be defined soon).

For $z \in \mathbf{F}$ and $m$ a positive integer, we define $z^m$ to denote the product of $z$ with itself $m$ times:

$$z^m = \underbrace{z \cdots \cdots z}_{m \text{ times}}.$$

Clearly $(z^m)^n = z^{mn}$ and $(wz)^m = w^m z^m$ for all $w, z \in \mathbf{F}$ and all positive integers $m, n$.

*The letter **F** is used because **R** and **C** are examples of what are called **fields**. In this book we will not need to deal with fields other than **R** or **C**. Many of the definitions, theorems, and proofs in linear algebra that work for both **R** and **C** also work without change if an arbitrary field replaces **R** or **C**.*

# Definition of Vector Space

Before defining what a vector space is, let's look at two important examples. The vector space $\mathbf{R}^2$, which you can think of as a plane, consists of all ordered pairs of real numbers:

$$\mathbf{R}^2 = \{(x, y) : x, y \in \mathbf{R}\}.$$

The vector space $\mathbf{R}^3$, which you can think of as ordinary space, consists of all ordered triples of real numbers:

$$\mathbf{R}^3 = \{(x, y, z) : x, y, z \in \mathbf{R}\}.$$

To generalize $\mathbf{R}^2$ and $\mathbf{R}^3$ to higher dimensions, we first need to discuss the concept of lists. Suppose $n$ is a nonnegative integer. A *list* of *length* $n$ is an ordered collection of $n$ objects (which might be numbers, other lists, or more abstract entities) separated by commas and surrounded by parentheses. A list of length $n$ looks like this:

$$(x_1, \ldots, x_n).$$

*Many mathematicians call a list of length $n$ an $n$-tuple.*

Thus a list of length 2 is an ordered pair and a list of length 3 is an ordered triple. For $j \in \{1, \ldots, n\}$, we say that $x_j$ is the $j^{\text{th}}$ *coordinate* of the list above. Thus $x_1$ is called the first coordinate, $x_2$ is called the second coordinate, and so on.

Sometimes we will use the word *list* without specifying its length. Remember, however, that by definition each list has a finite length that is a nonnegative integer, so that an object that looks like

$$(x_1, x_2, \ldots),$$

which might be said to have infinite length, is not a list. A list of length 0 looks like this: (). We consider such an object to be a list so that some of our theorems will not have trivial exceptions.

Two lists are equal if and only if they have the same length and the same coordinates in the same order. In other words, $(x_1, \ldots, x_m)$ equals $(y_1, \ldots, y_n)$ if and only if $m = n$ and $x_1 = y_1, \ldots, x_m = y_m$.

Lists differ from sets in two ways: in lists, order matters and repetitions are allowed, whereas in sets, order and repetitions are irrelevant. For example, the lists $(3, 5)$ and $(5, 3)$ are not equal, but the sets $\{3, 5\}$ and $\{5, 3\}$ are equal. The lists $(4, 4)$ and $(4, 4, 4)$ are not equal (they do not have the same length), though the sets $\{4,4\}$ and $\{4,4,4\}$ both equal the set $\{4\}$.

To define the higher-dimensional analogues of $\mathbf{R}^2$ and $\mathbf{R}^3$, we will simply replace $\mathbf{R}$ with $\mathbf{F}$ (which equals $\mathbf{R}$ or $\mathbf{C}$) and replace the 2 or 3 with an arbitrary positive integer. Specifically, fix a positive integer $n$ for the rest of this section. We define $\mathbf{F}^n$ to be the set of all lists of length $n$ consisting of elements of $\mathbf{F}$:

$$\mathbf{F}^n = \{(x_1, \ldots, x_n) : x_j \in \mathbf{F} \text{ for } j = 1, \ldots, n\}.$$

For example, if $\mathbf{F} = \mathbf{R}$ and $n$ equals 2 or 3, then this definition of $\mathbf{F}^n$ agrees with our previous notions of $\mathbf{R}^2$ and $\mathbf{R}^3$. As another example, $\mathbf{C}^4$ is the set of all lists of four complex numbers:

$$\mathbf{C}^4 = \{(z_1, z_2, z_3, z_4) : z_1, z_2, z_3, z_4 \in \mathbf{C}\}.$$

If $n \geq 4$, we cannot easily visualize $\mathbf{R}^n$ as a physical object. The same problem arises if we work with complex numbers: $\mathbf{C}^1$ can be thought of as a plane, but for $n \geq 2$, the human brain cannot provide geometric models of $\mathbf{C}^n$. However, even if $n$ is large, we can perform algebraic manipulations in $\mathbf{F}^n$ as easily as in $\mathbf{R}^2$ or $\mathbf{R}^3$. For example, addition is defined on $\mathbf{F}^n$ by adding corresponding coordinates:

**1.1**      $(x_1, \ldots, x_n) + (y_1, \ldots, y_n) = (x_1 + y_1, \ldots, x_n + y_n).$

*For an amusing account of how $\mathbf{R}^3$ would be perceived by a creature living in $\mathbf{R}^2$, read **Flatland: A Romance of Many Dimensions**, by Edwin A. Abbott. This novel, published in 1884, can help creatures living in three-dimensional space, such as ourselves, imagine a physical space of four or more dimensions.*

Often the mathematics of $\mathbf{F}^n$ becomes cleaner if we use a single entity to denote an list of $n$ numbers, without explicitly writing the coordinates. Thus the commutative property of addition on $\mathbf{F}^n$ should be expressed as

$$x + y = y + x$$

for all $x, y \in \mathbf{F}^n$, rather than the more cumbersome

$$(x_1, \ldots, x_n) + (y_1, \ldots, y_n) = (y_1, \ldots, y_n) + (x_1, \ldots, x_n)$$

for all $x_1, \ldots, x_n, y_1, \ldots, y_n \in \mathbf{F}$ (even though the latter formulation is needed to prove commutativity). If a single letter is used to denote an element of $\mathbf{F}^n$, then the same letter, with appropriate subscripts, is often used when coordinates must be displayed. For example, if $x \in \mathbf{F}^n$, then letting $x$ equal $(x_1, \ldots, x_n)$ is good notation. Even better, work with just $x$ and avoid explicit coordinates, if possible.

We let 0 denote the list of length $n$ all of whose coordinates are 0:

$$0 = (0, \ldots, 0).$$

Note that we are using the symbol 0 in two different ways—on the left side of the equation above, 0 denotes a list of length $n$, whereas on the right side, each 0 denotes a number. This potentially confusing practice actually causes no problems because the context always makes clear what is intended. For example, consider the statement that 0 is an additive identity for $\mathbf{F}^n$:

$$x + 0 = x$$

for all $x \in \mathbf{F}^n$. Here 0 must be a list because we have not defined the sum of an element of $\mathbf{F}^n$ (namely, $x$) and the number 0.

A picture can often aid our intuition. We will draw pictures depicting $\mathbf{R}^2$ because we can easily sketch this space on two-dimensional surfaces such as paper and blackboards. A typical element of $\mathbf{R}^2$ is a point $x = (x_1, x_2)$. Sometimes we think of $x$ not as a point but as an arrow starting at the origin and ending at $(x_1, x_2)$, as in the picture below. When we think of $x$ as an arrow, we refer to it as a ***vector***.

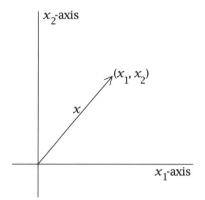

*Elements of $\mathbf{R}^2$ can be thought of as points or as vectors.*

The coordinate axes and the explicit coordinates unnecessarily clutter the picture above, and often you will gain better understanding by dispensing with them and just thinking of the vector, as in the next picture.

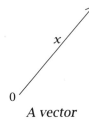

*A vector*

Whenever we use pictures in $\mathbf{R}^2$ or use the somewhat vague language of points and vectors, remember that these are just aids to our understanding, not substitutes for the actual mathematics that we will develop. Though we cannot draw good pictures in high-dimensional spaces, the elements of these spaces are as rigorously defined as elements of $\mathbf{R}^2$. For example, $(2, -3, 17, \pi, \sqrt{2})$ is an element of $\mathbf{R}^5$, and we may casually refer to it as a point in $\mathbf{R}^5$ or a vector in $\mathbf{R}^5$ without worrying about whether the geometry of $\mathbf{R}^5$ has any physical meaning.

Recall that we defined the sum of two elements of $\mathbf{F}^n$ to be the element of $\mathbf{F}^n$ obtained by adding corresponding coordinates; see 1.1. In the special case of $\mathbf{R}^2$, addition has a simple geometric interpretation. Suppose we have two vectors $x$ and $y$ in $\mathbf{R}^2$ that we want to add, as in the left side of the picture below. Move the vector $y$ parallel to itself so that its initial point coincides with the end point of the vector $x$. The sum $x + y$ then equals the vector whose initial point equals the initial point of $x$ and whose end point equals the end point of the moved vector $y$, as in the right side of the picture below.

*Mathematical models of the economy often have thousands of variables, say $x_1, \ldots, x_{5000}$, which means that we must operate in $\mathbf{R}^{5000}$. Such a space cannot be dealt with geometrically, but the algebraic approach works well. That's why our subject is called linear **algebra**.*

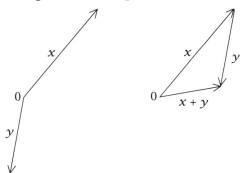

*The sum of two vectors*

Our treatment of the vector $y$ in the picture above illustrates a standard philosophy when we think of vectors in $\mathbf{R}^2$ as arrows: we can move an arrow parallel to itself (not changing its length or direction) and still think of it as the same vector.

Having dealt with addition in $\mathbf{F}^n$, we now turn to multiplication. We could define a multiplication on $\mathbf{F}^n$ in a similar fashion, starting with two elements of $\mathbf{F}^n$ and getting another element of $\mathbf{F}^n$ by multiplying corresponding coordinates. Experience shows that this definition is not useful for our purposes. Another type of multiplication, called scalar multiplication, will be central to our subject. Specifically, we need to define what it means to multiply an element of $\mathbf{F}^n$ by an element of $\mathbf{F}$. We make the obvious definition, performing the multiplication in each coordinate:

$$a(x_1,\ldots,x_n) = (ax_1,\ldots,ax_n);$$

here $a \in \mathbf{F}$ and $(x_1,\ldots,x_n) \in \mathbf{F}^n$.

*In scalar multiplication, we multiply together a scalar and a vector, getting a vector. You may be familiar with the dot product in $\mathbf{R}^2$ or $\mathbf{R}^3$, in which we multiply together two vectors and obtain a scalar. Generalizations of the dot product will become important when we study inner products in Chapter 6. You may also be familiar with the cross product in $\mathbf{R}^3$, in which we multiply together two vectors and obtain another vector. No useful generalization of this type of multiplication exists in higher dimensions.*

Scalar multiplication has a nice geometric interpretation in $\mathbf{R}^2$. If $a$ is a positive number and $x$ is a vector in $\mathbf{R}^2$, then $ax$ is the vector that points in the same direction as $x$ and whose length is $a$ times the length of $x$. In other words, to get $ax$, we shrink or stretch $x$ by a factor of $a$, depending upon whether $a < 1$ or $a > 1$. The next picture illustrates this point.

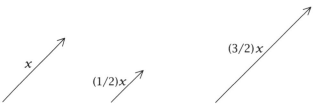

*Multiplication by positive scalars*

If $a$ is a negative number and $x$ is a vector in $\mathbf{R}^2$, then $ax$ is the vector that points in the opposite direction as $x$ and whose length is $|a|$ times the length of $x$, as illustrated in the next picture.

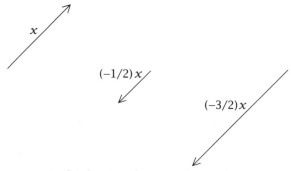

*Multiplication by negative scalars*

The motivation for the definition of a vector space comes from the important properties possessed by addition and scalar multiplication on $\mathbf{F}^n$. Specifically, addition on $\mathbf{F}^n$ is commutative and associative and has an identity, namely, 0. Every element has an additive inverse. Scalar multiplication on $\mathbf{F}^n$ is associative, and scalar multiplication by 1 acts as a multiplicative identity should. Finally, addition and scalar multiplication on $\mathbf{F}^n$ are connected by distributive properties.

We will define a vector space to be a set $V$ along with an addition and a scalar multiplication on $V$ that satisfy the properties discussed in the previous paragraph. By an **addition** on $V$ we mean a function that assigns an element $u + v \in V$ to each pair of elements $u, v \in V$. By a **scalar multiplication** on $V$ we mean a function that assigns an element $av \in V$ to each $a \in \mathbf{F}$ and each $v \in V$.

Now we are ready to give the formal definition of a vector space. A **vector space** is a set $V$ along with an addition on $V$ and a scalar multiplication on $V$ such that the following properties hold:

**commutativity**

$u + v = v + u$ for all $u, v \in V$;

**associativity**

$(u + v) + w = u + (v + w)$ and $(ab)v = a(bv)$ for all $u, v, w \in V$ and all $a, b \in \mathbf{F}$;

**additive identity**

there exists an element $0 \in V$ such that $v + 0 = v$ for all $v \in V$;

**additive inverse**

for every $v \in V$, there exists $w \in V$ such that $v + w = 0$;

**multiplicative identity**

$1v = v$ for all $v \in V$;

**distributive properties**

$a(u + v) = au + av$ and $(a + b)u = au + bu$ for all $a, b \in \mathbf{F}$ and all $u, v \in V$.

The scalar multiplication in a vector space depends upon $\mathbf{F}$. Thus when we need to be precise, we will say that $V$ is a vector space over $\mathbf{F}$ instead of saying simply that $V$ is a vector space. For example, $\mathbf{R}^n$ is a vector space over $\mathbf{R}$, and $\mathbf{C}^n$ is a vector space over $\mathbf{C}$. Frequently, a vector space over $\mathbf{R}$ is called a **real vector space** and a vector space over

**C** is called a ***complex vector space***. Usually the choice of **F** is either obvious from the context or irrelevant, and thus we often assume that **F** is lurking in the background without specifically mentioning it.

Elements of a vector space are called ***vectors*** or ***points***. This geometric language sometimes aids our intuition.

Not surprisingly, $\mathbf{F}^n$ is a vector space over **F**, as you should verify. Of course, this example motivated our definition of vector space.

*The simplest vector space contains only one point. In other words, {0} is a vector space, though not a very interesting one.*

For another example, consider $\mathbf{F}^\infty$, which is defined to be the set of all sequences of elements of **F**:

$$\mathbf{F}^\infty = \{(x_1, x_2, \dots) : x_j \in \mathbf{F} \text{ for } j = 1, 2, \dots\}.$$

Addition and scalar multiplication on $\mathbf{F}^\infty$ are defined as expected:

$$(x_1, x_2, \dots) + (y_1, y_2, \dots) = (x_1 + y_1, x_2 + y_2, \dots),$$
$$a(x_1, x_2, \dots) = (ax_1, ax_2, \dots).$$

With these definitions, $\mathbf{F}^\infty$ becomes a vector space over **F**, as you should verify. The additive identity in this vector space is the sequence consisting of all 0's.

Our next example of a vector space involves polynomials. A function $p: \mathbf{F} \to \mathbf{F}$ is called a ***polynomial*** with coefficients in **F** if there exist $a_0, \dots, a_m \in \mathbf{F}$ such that

$$p(z) = a_0 + a_1 z + a_2 z^2 + \dots + a_m z^m$$

*Though $\mathbf{F}^n$ is our crucial example of a vector space, not all vector spaces consist of lists. For example, the elements of $\mathcal{P}(\mathbf{F})$ consist of functions on **F**, not lists. In general, a vector space is an abstract entity whose elements might be lists, functions, or weird objects.*

for all $z \in \mathbf{F}$. We define $\mathcal{P}(\mathbf{F})$ to be the set of all polynomials with coefficients in **F**. Addition on $\mathcal{P}(\mathbf{F})$ is defined as you would expect: if $p, q \in \mathcal{P}(\mathbf{F})$, then $p + q$ is the polynomial defined by

$$(p + q)(z) = p(z) + q(z)$$

for $z \in \mathbf{F}$. For example, if $p$ is the polynomial defined by $p(z) = 2z + z^3$ and $q$ is the polynomial defined by $q(z) = 7 + 4z$, then $p + q$ is the polynomial defined by $(p + q)(z) = 7 + 6z + z^3$. Scalar multiplication on $\mathcal{P}(\mathbf{F})$ also has the obvious definition: if $a \in \mathbf{F}$ and $p \in \mathcal{P}(\mathbf{F})$, then $ap$ is the polynomial defined by

$$(ap)(z) = ap(z)$$

for $z \in \mathbf{F}$. With these definitions of addition and scalar multiplication, $\mathcal{P}(\mathbf{F})$ is a vector space, as you should verify. The additive identity in this vector space is the polynomial all of whose coefficients equal 0.

Soon we will see further examples of vector spaces, but first we need to develop some of the elementary properties of vector spaces.

# Properties of Vector Spaces

The definition of a vector space requires that it have an additive identity. The proposition below states that this identity is unique.

**1.2   Proposition:**   *A vector space has a unique additive identity.*

PROOF:  Suppose 0 and $0'$ are both additive identities for some vector space $V$. Then

$$0' = 0' + 0 = 0,$$

where the first equality holds because 0 is an additive identity and the second equality holds because $0'$ is an additive identity. Thus $0' = 0$, proving that $V$ has only one additive identity.   ■

*The symbol ■ means "end of the proof".*

Each element $v$ in a vector space has an additive inverse, an element $w$ in the vector space such that $v + w = 0$. The next proposition shows that each element in a vector space has only one additive inverse.

**1.3   Proposition:**   *Every element in a vector space has a unique additive inverse.*

PROOF:  Suppose $V$ is a vector space. Let $v \in V$. Suppose that $w$ and $w'$ are additive inverses of $v$. Then

$$w = w + 0 = w + (v + w') = (w + v) + w' = 0 + w' = w'.$$

Thus $w = w'$, as desired.   ■

Because additive inverses are unique, we can let $-v$ denote the additive inverse of a vector $v$. We define $w - v$ to mean $w + (-v)$.

Almost all the results in this book will involve some vector space. To avoid being distracted by having to restate frequently something such as "Assume that $V$ is a vector space", we now make the necessary declaration once and for all:

> Let's agree that for the rest of the book
> $V$ will denote a vector space over **F**.

Because of associativity, we can dispense with parentheses when dealing with additions involving more than two elements in a vector space. For example, we can write $u+v+w$ without parentheses because the two possible interpretations of that expression, namely, $(u+v)+w$ and $u + (v + w)$, are equal. We first use this familiar convention of not using parentheses in the next proof. In the next proposition, 0 denotes a scalar (the number $0 \in \mathbf{F}$) on the left side of the equation and a vector (the additive identity of $V$) on the right side of the equation.

*Note that 1.4 and 1.5 assert something about scalar multiplication and the additive identity of V. The only part of the definition of a vector space that connects scalar multiplication and vector addition is the distributive property. Thus the distributive property must be used in the proofs.*

**1.4    Proposition:**  $0v = 0$ *for every* $v \in V$.

PROOF:  For $v \in V$, we have

$$0v = (0 + 0)v = 0v + 0v.$$

Adding the additive inverse of $0v$ to both sides of the equation above gives $0 = 0v$, as desired.    ∎

In the next proposition, 0 denotes the additive identity of $V$. Though their proofs are similar, 1.4 and 1.5 are not identical. More precisely, 1.4 states that the product of the scalar 0 and any vector equals the vector 0, whereas 1.5 states that the product of any scalar and the vector 0 equals the vector 0.

**1.5    Proposition:**  $a0 = 0$ *for every* $a \in \mathbf{F}$.

PROOF:  For $a \in \mathbf{F}$, we have

$$a0 = a(0 + 0) = a0 + a0.$$

Adding the additive inverse of $a0$ to both sides of the equation above gives $0 = a0$, as desired.    ∎

Now we show that if an element of $V$ is multiplied by the scalar $-1$, then the result is the additive inverse of the element of $V$.

**1.6    Proposition:**  $(-1)v = -v$ *for every* $v \in V$.

PROOF:  For $v \in V$, we have

$$v + (-1)v = 1v + (-1)v = \big(1 + (-1)\big)v = 0v = 0.$$

This equation says that $(-1)v$, when added to $v$, gives 0. Thus $(-1)v$ must be the additive inverse of $v$, as desired.    ∎

# Subspaces

A subset $U$ of $V$ is called a **subspace** of $V$ if $U$ is also a vector space (using the same addition and scalar multiplication as on $V$). For example,

$$\{(x_1, x_2, 0) : x_1, x_2 \in \mathbf{F}\}$$

is a subspace of $\mathbf{F}^3$.

*Some mathematicians use the term **linear subspace**, which means the same as subspace.*

If $U$ is a subset of $V$, then to check that $U$ is a subspace of $V$ we need only check that $U$ satisfies the following:

**additive identity**
   $0 \in U$

**closed under addition**
   $u, v \in U$ implies $u + v \in U$;

**closed under scalar multiplication**
   $a \in \mathbf{F}$ and $u \in U$ implies $au \in U$.

The first condition insures that the additive identity of $V$ is in $U$. The second condition insures that addition makes sense on $U$. The third condition insures that scalar multiplication makes sense on $U$. To show that $U$ is a vector space, the other parts of the definition of a vector space do not need to be checked because they are automatically satisfied. For example, the associative and commutative properties of addition automatically hold on $U$ because they hold on the larger space $V$. As another example, if the third condition above holds and $u \in U$, then $-u$ (which equals $(-1)u$ by 1.6) is also in $U$, and hence every element of $U$ has an additive inverse in $U$.

*Clearly $\{0\}$ is the smallest subspace of $V$ and $V$ itself is the largest subspace of $V$. The empty set is not a subspace of $V$ because a subspace must be a vector space and a vector space must contain at least one element, namely, an additive identity.*

The three conditions above usually enable us to determine quickly whether a given subset of $V$ is a subspace of $V$. For example, if $b \in \mathbf{F}$, then

$$\{(x_1, x_2, x_3, x_4) \in \mathbf{F}^4 : x_3 = 5x_4 + b\}$$

is a subspace of $\mathbf{F}^4$ if and only if $b = 0$, as you should verify. As another example, you should verify that

$$\{p \in \mathcal{P}(\mathbf{F}) : p(3) = 0\}$$

is a subspace of $\mathcal{P}(\mathbf{F})$.

The subspaces of $\mathbf{R}^2$ are precisely $\{0\}$, $\mathbf{R}^2$, and all lines in $\mathbf{R}^2$ through the origin. The subspaces of $\mathbf{R}^3$ are precisely $\{0\}$, $\mathbf{R}^3$, all lines in $\mathbf{R}^3$

through the origin, and all planes in $\mathbf{R}^3$ through the origin. To prove that all these objects are indeed subspaces is easy—the hard part is to show that they are the only subspaces of $\mathbf{R}^2$ or $\mathbf{R}^3$. That task will be easier after we introduce some additional tools in the next chapter.

## Sums and Direct Sums

In later chapters, we will find that the notions of vector space sums and direct sums are useful. We define these concepts here.

*When dealing with vector spaces, we are usually interested only in subspaces, as opposed to arbitrary subsets. The union of subspaces is rarely a subspace (see Exercise 9 in this chapter), which is why we usually work with sums rather than unions.*

Suppose $U_1, \ldots, U_m$ are subspaces of $V$. The **sum** of $U_1, \ldots, U_m$, denoted $U_1 + \cdots + U_m$, is defined to be the set of all possible sums of elements of $U_1, \ldots, U_m$. More precisely,

$$U_1 + \cdots + U_m = \{u_1 + \cdots + u_m : u_1 \in U_1, \ldots, u_m \in U_m\}.$$

You should verify that if $U_1, \ldots, U_m$ are subspaces of $V$, then the sum $U_1 + \cdots + U_m$ is a subspace of $V$.

Let's look at some examples of sums of subspaces. Suppose $U$ is the set of all elements of $\mathbf{F}^3$ whose second and third coordinates equal 0, and $W$ is the set of all elements of $\mathbf{F}^3$ whose first and third coordinates equal 0:

$$U = \{(x, 0, 0) \in \mathbf{F}^3 : x \in \mathbf{F}\} \quad \text{and} \quad W = \{(0, y, 0) \in \mathbf{F}^3 : y \in \mathbf{F}\}.$$

Then

*Sums of subspaces in the theory of vector spaces are analogous to unions of subsets in set theory. Given two subspaces of a vector space, the smallest subspace containing them is their sum. Analogously, given two subsets of a set, the smallest subset containing them is their union.*

**1.7**          $$U + W = \{(x, y, 0) : x, y \in \mathbf{F}\},$$

as you should verify.

As another example, suppose $U$ is as above and $W$ is the set of all elements of $\mathbf{F}^3$ whose first and second coordinates equal each other and whose third coordinate equals 0:

$$W = \{(y, y, 0) \in \mathbf{F}^3 : y \in \mathbf{F}\}.$$

Then $U + W$ is also given by 1.7, as you should verify.

Suppose $U_1, \ldots, U_m$ are subspaces of $V$. Clearly $U_1, \ldots, U_m$ are all contained in $U_1 + \cdots + U_m$ (to see this, consider sums $u_1 + \cdots + u_m$ where all except one of the $u$'s are 0). Conversely, any subspace of $V$ containing $U_1, \ldots, U_m$ must contain $U_1 + \cdots + U_m$ (because subspaces must contain all finite sums of their elements). Thus $U_1 + \cdots + U_m$ is the smallest subspace of $V$ containing $U_1, \ldots, U_m$.

Suppose $U_1, \ldots, U_m$ are subspaces of $V$ such that $V = U_1 + \cdots + U_m$. Thus every element of $V$ can be written in the form

$$u_1 + \cdots + u_m,$$

where each $u_j \in U_j$. We will be especially interested in cases where each vector in $V$ can be uniquely represented in the form above. This situation is so important that we give it a special name: direct sum. Specifically, we say that $V$ is the **direct sum** of subspaces $U_1, \ldots, U_m$, written $V = U_1 \oplus \cdots \oplus U_m$, if each element of $V$ can be written uniquely as a sum $u_1 + \cdots + u_m$, where each $u_j \in U_j$.

Let's look at some examples of direct sums. Suppose $U$ is the subspace of $\mathbf{F}^3$ consisting of those vectors whose last coordinate equals 0, and $W$ is the subspace of $\mathbf{F}^3$ consisting of those vectors whose first two coordinates equal 0:

$$U = \{(x, y, 0) \in \mathbf{F}^3 : x, y \in \mathbf{F}\} \quad \text{and} \quad W = \{(0, 0, z) \in \mathbf{F}^3 : z \in \mathbf{F}\}.$$

Then $\mathbf{F}^3 = U \oplus W$, as you should verify.

As another example, suppose $U_j$ is the subspace of $\mathbf{F}^n$ consisting of those vectors whose coordinates are all 0, except possibly in the $j^{\text{th}}$ slot (for example, $U_2 = \{(0, x, 0, \ldots, 0) \in \mathbf{F}^n : x \in \mathbf{F}\}$). Then

$$\mathbf{F}^n = U_1 \oplus \cdots \oplus U_n,$$

as you should verify.

As a final example, consider the vector space $\mathcal{P}(\mathbf{F})$ of all polynomials with coefficients in $\mathbf{F}$. Let $U_e$ denote the subspace of $\mathcal{P}(\mathbf{F})$ consisting of all polynomials $p$ of the form

$$p(z) = a_0 + a_2 z^2 + \cdots + a_{2m} z^{2m},$$

and let $U_o$ denote the subspace of $\mathcal{P}(\mathbf{F})$ consisting of all polynomials $p$ of the form

$$p(z) = a_1 z + a_3 z^3 + \cdots + a_{2m+1} z^{2m+1};$$

here $m$ is a nonnegative integer and $a_0, \ldots, a_{2m+1} \in \mathbf{F}$ (the notations $U_e$ and $U_o$ should remind you of even and odd powers of $z$). You should verify that

*The symbol $\oplus$, consisting of a plus sign inside a circle, is used to denote direct sums as a reminder that we are dealing with a special type of sum of subspaces—each element in the direct sum can be represented only one way as a sum of elements from the specified subspaces.*

$$\mathcal{P}(\mathbf{F}) = U_e \oplus U_o.$$

Sometimes nonexamples add to our understanding as much as examples. Consider the following three subspaces of $\mathbf{F}^3$:

$$U_1 = \{(x, y, 0) \in \mathbf{F}^3 : x, y \in \mathbf{F}\};$$
$$U_2 = \{(0, 0, z) \in \mathbf{F}^3 : z \in \mathbf{F}\};$$
$$U_3 = \{(0, y, y) \in \mathbf{F}^3 : y \in \mathbf{F}\}.$$

Clearly $\mathbf{F}^3 = U_1 + U_2 + U_3$ because an arbitrary vector $(x, y, z) \in \mathbf{F}^3$ can be written as

$$(x, y, z) = (x, y, 0) + (0, 0, z) + (0, 0, 0),$$

where the first vector on the right side is in $U_1$, the second vector is in $U_2$, and the third vector is in $U_3$. However, $\mathbf{F}^3$ does not equal the direct sum of $U_1, U_2, U_3$ because the vector $(0, 0, 0)$ can be written in two different ways as a sum $u_1 + u_2 + u_3$, with each $u_j \in U_j$. Specifically, we have

$$(0, 0, 0) = (0, 1, 0) + (0, 0, 1) + (0, -1, -1)$$

and, of course,

$$(0, 0, 0) = (0, 0, 0) + (0, 0, 0) + (0, 0, 0),$$

where the first vector on the right side of each equation above is in $U_1$, the second vector is in $U_2$, and the third vector is in $U_3$.

In the example above, we showed that something is not a direct sum by showing that 0 does not have a unique representation as a sum of appropriate vectors. The definition of direct sum requires that every vector in the space have a unique representation as an appropriate sum. Suppose we have a collection of subspaces whose sum equals the whole space. The next proposition shows that when deciding whether this collection of subspaces is a direct sum, we need only consider whether 0 can be uniquely written as an appropriate sum.

**1.8**    **Proposition:** *Suppose that $U_1, \ldots, U_n$ are subspaces of $V$. Then $V = U_1 \oplus \cdots \oplus U_n$ if and only if both the following conditions hold:*

(a)      $V = U_1 + \cdots + U_n$;

(b)      *the only way to write 0 as a sum $u_1 + \cdots + u_n$, where each $u_j \in U_j$, is by taking all the $u_j$'s equal to 0.*

PROOF:   First suppose that $V = U_1 \oplus \cdots \oplus U_n$. Clearly (a) holds (because of how sum and direct sum are defined). To prove (b), suppose that $u_1 \in U_1, \ldots, u_n \in U_n$ and

$$0 = u_1 + \cdots + u_n.$$

Then each $u_j$ must be 0 (this follows from the uniqueness part of the definition of direct sum because $0 = 0 + \cdots + 0$ and $0 \in U_1, \ldots, 0 \in U_n$), proving (b).

Now suppose that (a) and (b) hold. Let $v \in V$. By (a), we can write

$$v = u_1 + \cdots + u_n$$

for some $u_1 \in U_1, \ldots, u_n \in U_n$. To show that this representation is unique, suppose that we also have

$$v = v_1 + \cdots + v_n,$$

where $v_1 \in U_1, \ldots, v_n \in U_n$. Subtracting these two equations, we have

$$0 = (u_1 - v_1) + \cdots + (u_n - v_n).$$

Clearly $u_1 - v_1 \in U_1, \ldots, u_n - v_n \in U_n$, so the equation above and (b) imply that each $u_j - v_j = 0$. Thus $u_1 = v_1, \ldots, u_n = v_n$, as desired. ∎

The next proposition gives a simple condition for testing which pairs of subspaces give a direct sum. Note that this proposition deals only with the case of two subspaces. When asking about a possible direct sum with more than two subspaces, it is not enough to test that any two of the subspaces intersect only at 0. To see this, consider the nonexample presented just before 1.8. In that nonexample, we had $\mathbf{F}^3 = U_1 + U_2 + U_3$, but $\mathbf{F}^3$ did not equal the direct sum of $U_1, U_2, U_3$. However, in that nonexample, we have $U_1 \cap U_2 = U_1 \cap U_3 = U_2 \cap U_3 = \{0\}$ (as you should verify). The next proposition shows that with just two subspaces we get a nice necessary and sufficient condition for a direct sum.

*Sums of subspaces are analogous to unions of subsets. Similarly, direct sums of subspaces are analogous to disjoint unions of subsets. No two subspaces of a vector space can be disjoint because both must contain 0. So disjointness is replaced, at least in the case of two subspaces, with the requirement that the intersection equals $\{0\}$.*

**1.9   Proposition:** *Suppose that $U$ and $W$ are subspaces of $V$. Then $V = U \oplus W$ if and only if $V = U + W$ and $U \cap W = \{0\}$.*

PROOF:   First suppose that $V = U \oplus W$. Then $V = U + W$ (by the definition of direct sum). Also, if $v \in U \cap W$, then $0 = v + (-v)$, where

$v \in U$ and $-v \in W$. By the unique representation of 0 as the sum of a vector in $U$ and a vector in $W$, we must have $v = 0$. Thus $U \cap W = \{0\}$, completing the proof in one direction.

To prove the other direction, now suppose that $V = U + W$ and $U \cap W = \{0\}$. To prove that $V = U \oplus W$, suppose that

$$0 = u + w,$$

where $u \in U$ and $w \in W$. To complete the proof, we need only show that $u = w = 0$ (by 1.8). The equation above implies that $u = -w \in W$. Thus $u \in U \cap W$, and hence $u = 0$. This, along with equation above, implies that $w = 0$, completing the proof.   ∎

# Exercises

1.  Suppose $a$ and $b$ are real numbers, not both 0. Find real numbers $c$ and $d$ such that
    $$1/(a + bi) = c + di.$$

2.  Show that
    $$\frac{-1 + \sqrt{3}i}{2}$$
    is a cube root of 1 (meaning that its cube equals 1).

3.  Prove that $-(-v) = v$ for every $v \in V$.

4.  Prove that if $a \in \mathbf{F}$, $v \in V$, and $av = 0$, then $a = 0$ or $v = 0$.

5.  For each of the following subsets of $\mathbf{F}^3$, determine whether it is a subspace of $\mathbf{F}^3$:

    (a)  $\{(x_1, x_2, x_3) \in \mathbf{F}^3 : x_1 + 2x_2 + 3x_3 = 0\}$;

    (b)  $\{(x_1, x_2, x_3) \in \mathbf{F}^3 : x_1 + 2x_2 + 3x_3 = 4\}$;

    (c)  $\{(x_1, x_2, x_3) \in \mathbf{F}^3 : x_1 x_2 x_3 = 0\}$;

    (d)  $\{(x_1, x_2, x_3) \in \mathbf{F}^3 : x_1 = 5x_3\}$.

6.  Give an example of a nonempty subset $U$ of $\mathbf{R}^2$ such that $U$ is closed under addition and under taking additive inverses (meaning $-u \in U$ whenever $u \in U$), but $U$ is not a subspace of $\mathbf{R}^2$.

7.  Give an example of a nonempty subset $U$ of $\mathbf{R}^2$ such that $U$ is closed under scalar multiplication, but $U$ is not a subspace of $\mathbf{R}^2$.

8.  Prove that the intersection of any collection of subspaces of $V$ is a subspace of $V$.

9.  Prove that the union of two subspaces of $V$ is a subspace of $V$ if and only if one of the subspaces is contained in the other.

10. Suppose that $U$ is a subspace of $V$. What is $U + U$?

11. Is the operation of addition on the subspaces of $V$ commutative? Associative? (In other words, if $U_1, U_2, U_3$ are subspaces of $V$, is $U_1 + U_2 = U_2 + U_1$? Is $(U_1 + U_2) + U_3 = U_1 + (U_2 + U_3)$?)

12.    Does the operation of addition on the subspaces of $V$ have an additive identity? Which subspaces have additive inverses?

13.    Prove or give a counterexample: if $U_1, U_2, W$ are subspaces of $V$ such that
$$U_1 + W = U_2 + W,$$
then $U_1 = U_2$.

14.    Suppose $U$ is the subspace of $\mathcal{P}(\mathbf{F})$ consisting of all polynomials $p$ of the form
$$p(z) = az^2 + bz^5,$$
where $a, b \in \mathbf{F}$.  Find a subspace $W$ of $\mathcal{P}(\mathbf{F})$ such that $\mathcal{P}(\mathbf{F}) = U \oplus W$.

15.    Prove or give a counterexample: if $U_1, U_2, W$ are subspaces of $V$ such that
$$V = U_1 \oplus W \quad \text{and} \quad V = U_2 \oplus W,$$
then $U_1 = U_2$.

# CHAPTER 2

# *Finite-Dimensional Vector Spaces*

In the last chapter we learned about vector spaces. Linear algebra focuses not on arbitrary vector spaces, but on finite-dimensional vector spaces, which we introduce in this chapter. Here we will deal with the key concepts associated with these spaces: span, linear independence, basis, and dimension.

Let's review our standing assumptions:

> Recall that **F** denotes **R** or **C**.
> Recall also that $V$ is a vector space over **F**.

# Span and Linear Independence

A **linear combination** of a list $(v_1, \ldots, v_m)$ of vectors in $V$ is a vector of the form

**2.1**
$$a_1 v_1 + \cdots + a_m v_m,$$

*Some mathematicians use the term **linear span**, which means the same as span.*

where $a_1, \ldots, a_m \in \mathbf{F}$. The set of all linear combinations of $(v_1, \ldots, v_m)$ is called the **span** of $(v_1, \ldots, v_m)$, denoted span$(v_1, \ldots, v_m)$. In other words,

$$\text{span}(v_1, \ldots, v_m) = \{a_1 v_1 + \cdots + a_m v_m : a_1, \ldots, a_m \in \mathbf{F}\}.$$

As an example of these concepts, suppose $V = \mathbf{F}^3$. The vector $(7, 2, 9)$ is a linear combination of $((2, 1, 3), (1, 0, 1))$ because

$$(7, 2, 9) = 2(2, 1, 3) + 3(1, 0, 1).$$

Thus $(7, 2, 9) \in \text{span}((2, 1, 3), (1, 0, 1))$.

You should verify that the span of any list of vectors in $V$ is a subspace of $V$. To be consistent, we declare that the span of the empty list () equals $\{0\}$ (recall that the empty set is not a subspace of $V$).

If $(v_1, \ldots, v_m)$ is a list of vectors in $V$, then each $v_j$ is a linear combination of $(v_1, \ldots, v_m)$ (to show this, set $a_j = 1$ and let the other $a$'s in 2.1 equal 0). Thus span$(v_1, \ldots, v_m)$ contains each $v_j$. Conversely, because subspaces are closed under scalar multiplication and addition, every subspace of $V$ containing each $v_j$ must contain span$(v_1, \ldots, v_m)$. Thus the span of a list of vectors in $V$ is the smallest subspace of $V$ containing all the vectors in the list.

*Recall that by definition every list has finite length.*

If span$(v_1, \ldots, v_m)$ equals $V$, we say that $(v_1, \ldots, v_m)$ **spans** $V$. A vector space is called **finite dimensional** if some list of vectors in it spans the space. For example, $\mathbf{F}^n$ is finite dimensional because

$$((1, 0, \ldots, 0), (0, 1, 0, \ldots, 0), \ldots, (0, \ldots, 0, 1))$$

spans $\mathbf{F}^n$, as you should verify.

Before giving the next example of a finite-dimensional vector space, we need to define the degree of a polynomial. A polynomial $p \in \mathcal{P}(\mathbf{F})$ is said to have **degree** $m$ if there exist scalars $a_0, a_1, \ldots, a_m \in \mathbf{F}$ with $a_m \neq 0$ such that

**2.2**
$$p(z) = a_0 + a_1 z + \cdots + a_m z^m$$

for all $z \in \mathbf{F}$. The polynomial that is identically 0 is said to have degree $-\infty$.

For $m$ a nonnegative integer, let $\mathcal{P}_m(\mathbf{F})$ denote the set of all polynomials with coefficients in $\mathbf{F}$ and degree at most $m$. You should verify that $\mathcal{P}_m(\mathbf{F})$ is a subspace of $\mathcal{P}(\mathbf{F})$; hence $\mathcal{P}_m(\mathbf{F})$ is a vector space. This vector space is finite dimensional because it is spanned by the list $(1, z, \ldots, z^m)$; here we are slightly abusing notation by letting $z^k$ denote a function (so $z$ is a dummy variable).

A vector space that is not finite dimensional is called ***infinite dimensional***. For example, $\mathcal{P}(\mathbf{F})$ is infinite dimensional. To prove this, consider any list of elements of $\mathcal{P}(\mathbf{F})$. Let $m$ denote the highest degree of any of the polynomials in the list under consideration (recall that by definition a list has finite length). Then every polynomial in the span of this list must have degree at most $m$. Thus our list cannot span $\mathcal{P}(\mathbf{F})$. Because no list spans $\mathcal{P}(\mathbf{F})$, this vector space is infinite dimensional.

The vector space $\mathbf{F}^\infty$, consisting of all sequences of elements of $\mathbf{F}$, is also infinite dimensional, though this is a bit harder to prove. You should be able to give a proof by using some of the tools we will soon develop.

Suppose $v_1, \ldots, v_m \in V$ and $v \in \operatorname{span}(v_1, \ldots, v_m)$. By the definition of span, there exist $a_1, \ldots, a_m \in \mathbf{F}$ such that

$$v = a_1 v_1 + \cdots + a_m v_m.$$

Consider the question of whether the choice of $a$'s in the equation above is unique. Suppose $\hat{a}_1, \ldots, \hat{a}_m$ is another set of scalars such that

$$v = \hat{a}_1 v_1 + \cdots + \hat{a}_m v_m.$$

Subtracting the last two equations, we have

$$0 = (a_1 - \hat{a}_1)v_1 + \cdots + (a_m - \hat{a}_m)v_m.$$

Thus we have written 0 as a linear combination of $(v_1, \ldots, v_m)$. If the only way to do this is the obvious way (using 0 for all scalars), then each $a_j - \hat{a}_j$ equals 0, which means that each $a_j$ equals $\hat{a}_j$ (and thus the choice of $a$'s was indeed unique). This situation is so important that we give it a special name—linear independence—which we now define.

A list $(v_1, \ldots, v_m)$ of vectors in $V$ is called ***linearly independent*** if the only choice of $a_1, \ldots, a_m \in \mathbf{F}$ that makes $a_1 v_1 + \cdots + a_m v_m$ equal 0 is $a_1 = \cdots = a_m = 0$. For example,

*Infinite-dimensional vector spaces, which we will not mention much anymore, are the center of attention in the branch of mathematics called **functional analysis**. Functional analysis uses tools from both analysis and algebra.*

$$((1,0,0,0),(0,1,0,0),(0,0,1,0))$$

is linearly independent in $\mathbf{F}^4$, as you should verify. The reasoning in the previous paragraph shows that $(v_1,\ldots,v_m)$ is linearly independent if and only if each vector in span$(v_1,\ldots,v_m)$ has only one representation as a linear combination of $(v_1,\ldots,v_m)$.

*Most linear algebra texts define linearly independent sets instead of linearly independent lists. With that definition, the set $\{(0,1),(0,1),(1,0)\}$ is linearly independent in $\mathbf{F}^2$ because it equals the set $\{(0,1),(1,0)\}$. With our definition, the list $\big((0,1),(0,1),(1,0)\big)$ is not linearly independent (because 1 times the first vector plus $-1$ times the second vector plus 0 times the third vector equals 0). By dealing with lists instead of sets, we will avoid some problems associated with the usual approach.*

For another example of a linearly independent list, fix a nonnegative integer $m$. Then $(1,z,\ldots,z^m)$ is linearly independent in $\mathcal{P}(\mathbf{F})$. To verify this, suppose that $a_0,a_1,\ldots,a_m \in \mathbf{F}$ are such that

**2.3**          $$a_0 + a_1 z + \cdots + a_m z^m = 0$$

for every $z \in \mathbf{F}$. If at least one of the coefficients $a_0,a_1,\ldots,a_m$ were nonzero, then 2.3 could be satisfied by at most $m$ distinct values of $z$ (if you are unfamiliar with this fact, just believe it for now; we will prove it in Chapter 4); this contradiction shows that all the coefficients in 2.3 equal 0. Hence $(1,z,\ldots,z^m)$ is linearly independent, as claimed.

A list of vectors in $V$ is called **linearly dependent** if it is not linearly independent. In other words, a list $(v_1,\ldots,v_m)$ of vectors in $V$ is linearly dependent if there exist $a_1,\ldots,a_m \in \mathbf{F}$, not all 0, such that $a_1 v_1 + \cdots + a_m v_m = 0$. For example, $((2,3,1),(1,-1,2),(7,3,8))$ is linearly dependent in $\mathbf{F}^3$ because

$$2(2,3,1) + 3(1,-1,2) + (-1)(7,3,8) = (0,0,0).$$

As another example, any list of vectors containing the 0 vector is linearly dependent (why?).

You should verify that a list $(v)$ of length 1 is linearly independent if and only if $v \neq 0$. You should also verify that a list of length 2 is linearly independent if and only if neither vector is a scalar multiple of the other. Caution: a list of length three or more may be linearly dependent even though no vector in the list is a scalar multiple of any other vector in the list, as shown by the example in the previous paragraph.

If some vectors are removed from a linearly independent list, the remaining list is also linearly independent, as you should verify. To allow this to remain true even if we remove all the vectors, we declare the empty list () to be linearly independent.

The lemma below will often be useful. It states that given a linearly dependent list of vectors, with the first vector not zero, one of the vectors is in the span of the previous ones and furthermore we can throw out that vector without changing the span of the original list.

**2.4    Linear Dependence Lemma:** *If $(v_1, \ldots, v_m)$ is linearly dependent in V and $v_1 \neq 0$, then there exists $j \in \{2, \ldots, m\}$ such that the following hold:*

(a)    $v_j \in \text{span}(v_1, \ldots, v_{j-1})$;

(b)    *if the $j^{th}$ term is removed from $(v_1, \ldots, v_m)$, the span of the remaining list equals $\text{span}(v_1, \ldots, v_m)$.*

PROOF: Suppose $(v_1, \ldots, v_m)$ is linearly dependent in $V$ and $v_1 \neq 0$. Then there exist $a_1, \ldots, a_m \in \mathbf{F}$, not all 0, such that

$$a_1 v_1 + \cdots + a_m v_m = 0.$$

Not all of $a_2, a_3, \ldots, a_m$ can be 0 (because $v_1 \neq 0$). Let $j$ be the largest element of $\{2, \ldots, m\}$ such that $a_j \neq 0$. Then

**2.5**                 $$v_j = -\frac{a_1}{a_j} v_1 - \cdots - \frac{a_{j-1}}{a_j} v_{j-1},$$

proving (a).

To prove (b), suppose that $u \in \text{span}(v_1, \ldots, v_m)$. Then there exist $c_1, \ldots, c_m \in \mathbf{F}$ such that

$$u = c_1 v_1 + \cdots + c_m v_m.$$

In the equation above, we can replace $v_j$ with the right side of 2.5, which shows that $u$ is in the span of the list obtained by removing the $j^{th}$ term from $(v_1, \ldots, v_m)$. Thus (b) holds.    ∎

Now we come to a key result. It says that linearly independent lists are never longer than spanning lists.

**2.6    Theorem:** *In a finite-dimensional vector space, the length of every linearly independent list of vectors is less than or equal to the length of every spanning list of vectors.*

PROOF: Suppose that $(u_1, \ldots, u_m)$ is linearly independent in $V$ and that $(w_1, \ldots, w_n)$ spans $V$. We need to prove that $m \leq n$. We do so through the multistep process described below; note that in each step we add one of the $u$'s and remove one of the $w$'s.

*Suppose that for each positive integer $m$, there exists a linearly independent list of $m$ vectors in $V$. Then this theorem implies that $V$ is infinite dimensional.*

**Step 1**

The list $(w_1, \ldots, w_n)$ spans $V$, and thus adjoining any vector to it produces a linearly dependent list. In particular, the list

$$(u_1, w_1, \ldots, w_n)$$

is linearly dependent. Thus by the linear dependence lemma (2.4), we can remove one of the $w$'s so that the list $B$ (of length $n$) consisting of $u_1$ and the remaining $w$'s spans $V$.

**Step j**

The list $B$ (of length $n$) from step $j-1$ spans $V$, and thus adjoining any vector to it produces a linearly dependent list. In particular, the list of length $(n+1)$ obtained by adjoining $u_j$ to $B$, placing it just after $u_1, \ldots, u_{j-1}$, is linearly dependent. By the linear dependence lemma (2.4), one of the vectors in this list is in the span of the previous ones, and because $(u_1, \ldots, u_j)$ is linearly independent, this vector must be one of the $w$'s, not one of the $u$'s. We can remove that $w$ from $B$ so that the new list $B$ (of length $n$) consisting of $u_1, \ldots, u_j$ and the remaining $w$'s spans $V$.

After step $m$, we have added all the $u$'s and the process stops. If at any step we added a $u$ and had no more $w$'s to remove, then we would have a contradiction. Thus there must be at least as many $w$'s as $u$'s. ∎

Our intuition tells us that any vector space contained in a finite-dimensional vector space should also be finite dimensional. We now prove that this intuition is correct.

**2.7    Proposition:**  *Every subspace of a finite-dimensional vector space is finite dimensional.*

PROOF:  Suppose $V$ is finite dimensional and $U$ is a subspace of $V$. We need to prove that $U$ is finite dimensional. We do this through the following multistep construction.

**Step 1**

If $U = \{0\}$, then $U$ is finite dimensional and we are done. If $U \neq \{0\}$, then choose a nonzero vector $v_1 \in U$.

**Step j**

If $U = \text{span}(v_1, \ldots, v_{j-1})$, then $U$ is finite dimensional and we are done. If $U \neq \text{span}(v_1, \dots, v_{j-1})$, then choose a vector $v_j \in U$ such that

$$v_j \notin \text{span}(v_1, \dots, v_{j-1}).$$

After each step, as long as the process continues, we have constructed a list of vectors such that no vector in this list is in the span of the previous vectors. Thus after each step we have constructed a linearly independent list, by the linear dependence lemma (2.4). This linearly independent list cannot be longer than any spanning list of $V$ (by 2.6), and thus the process must eventually terminate, which means that $U$ is finite dimensional. ∎

# Bases

A **basis** of $V$ is a list of vectors in $V$ that is linearly independent and spans $V$. For example,

$$((1, 0, \dots, 0), (0, 1, 0, \dots, 0), \dots, (0, \dots, 0, 1))$$

is a basis of $\mathbf{F}^n$, called the **standard basis** of $\mathbf{F}^n$. In addition to the standard basis, $\mathbf{F}^n$ has many other bases. For example, $((1, 2), (3, 5))$ is a basis of $\mathbf{F}^2$. The list $((1, 2))$ is linearly independent but is not a basis of $\mathbf{F}^2$ because it does not span $\mathbf{F}^2$. The list $((1, 2), (3, 5), (4, 7))$ spans $\mathbf{F}^2$ but is not a basis because it is not linearly independent. As another example, $(1, z, \dots, z^m)$ is a basis of $\mathcal{P}_m(\mathbf{F})$.

The next proposition helps explain why bases are useful.

**2.8    Proposition:** *A list $(v_1, \dots, v_n)$ of vectors in $V$ is a basis of $V$ if and only if every $v \in V$ can be written uniquely in the form*

**2.9**                    $$v = a_1 v_1 + \cdots + a_n v_n,$$

*where $a_1, \dots, a_n \in \mathbf{F}$.*

PROOF: First suppose that $(v_1, \dots, v_n)$ is a basis of $V$. Let $v \in V$. Because $(v_1, \dots, v_n)$ spans $V$, there exist $a_1, \dots, a_n \in \mathbf{F}$ such that 2.9 holds. To show that the representation in 2.9 is unique, suppose that $b_1, \dots, b_n$ are scalars so that we also have

$$v = b_1 v_1 + \cdots + b_n v_n.$$

*This proof is essentially a repetition of the ideas that led us to the definition of linear independence.*

Subtracting the last equation from 2.9, we get

$$0 = (a_1 - b_1)v_1 + \cdots + (a_n - b_n)v_n.$$

This implies that each $a_j - b_j = 0$ (because $(v_1, \ldots, v_n)$ is linearly independent) and hence $a_1 = b_1, \ldots, a_n = b_n$. We have the desired uniqueness, completing the proof in one direction.

For the other direction, suppose that every $v \in V$ can be written uniquely in the form given by 2.9. Clearly this implies that $(v_1, \ldots, v_n)$ spans $V$. To show that $(v_1, \ldots, v_n)$ is linearly independent, suppose that $a_1, \ldots, a_n \in \mathbf{F}$ are such that

$$0 = a_1 v_1 + \cdots + a_n v_n.$$

The uniqueness of the representation 2.9 (with $v = 0$) implies that $a_1 = \cdots = a_n = 0$. Thus $(v_1, \ldots, v_n)$ is linearly independent and hence is a basis of $V$. ∎

A spanning list in a vector space may not be a basis because it is not linearly independent. Our next result says that given any spanning list, some of the vectors in it can be discarded so that the remaining list is linearly independent and still spans the vector space.

**2.10　Theorem:** *Every spanning list in a vector space can be reduced to a basis of the vector space.*

PROOF:　Suppose $(v_1, \ldots, v_n)$ spans $V$. We want to remove some of the vectors from $(v_1, \ldots, v_n)$ so that the remaining vectors form a basis of $V$. We do this through the multistep process described below. Start with $B = (v_1, \ldots, v_n)$.

**Step 1**
　　If $v_1 = 0$, delete $v_1$ from $B$. If $v_1 \neq 0$, leave $B$ unchanged.

**Step j**
　　If $v_j$ is in span$(v_1, \ldots, v_{j-1})$, delete $v_j$ from $B$. If $v_j$ is not in span$(v_1, \ldots, v_{j-1})$, leave $B$ unchanged.

Stop the process after step $n$, getting a list $B$. This list $B$ spans $V$ because our original list spanned $B$ and we have discarded only vectors that were already in the span of the previous vectors. The process insures that no vector in $B$ is in the span of the previous ones. Thus $B$ is linearly independent, by the linear dependence lemma (2.4). Hence $B$ is a basis of $V$.                                                          ■

Consider the list

$$((1,2),(3,6),(4,7),(5,9)),$$

which spans $\mathbf{F}^2$. To make sure that you understand the last proof, you should verify that the process in the proof produces $((1,2),(4,7))$, a basis of $\mathbf{F}^2$, when applied to the list above.

Our next result, an easy corollary of the last theorem, tells us that every finite-dimensional vector space has a basis.

**2.11   Corollary:** *Every finite-dimensional vector space has a basis.*

PROOF:  By definition, a finite-dimensional vector space has a spanning list. The previous theorem tells us that any spanning list can be reduced to a basis.                                                          ■

We have crafted our definitions so that the finite-dimensional vector space $\{0\}$ is not a counterexample to the corollary above. In particular, the empty list () is a basis of the vector space $\{0\}$ because this list has been defined to be linearly independent and to have span $\{0\}$.

Our next theorem is in some sense a dual of 2.10, which said that every spanning list can be reduced to a basis. Now we show that given any linearly independent list, we can adjoin some additional vectors so that the extended list is still linearly independent but also spans the space.

**2.12   Theorem:** *Every linearly independent list of vectors in a finite-dimensional vector space can be extended to a basis of the vector space.*

PROOF:  Suppose $V$ is finite dimensional and $(v_1, \ldots, v_m)$ is linearly independent in $V$. We want to extend $(v_1, \ldots, v_m)$ to a basis of $V$. We do this through the multistep process described below. First we let $(w_1, \ldots, w_n)$ be any list of vectors in $V$ that spans $V$.

**Step 1**
> If $w_1$ is in the span of $(v_1, \ldots, v_m)$, let $B = (v_1, \ldots, v_m)$. If $w_1$ is not in the span of $(v_1, \ldots, v_m)$, let $B = (v_1, \ldots, v_m, w_1)$.

*This theorem can be used to give another proof of the previous corollary. Specifically, suppose $V$ is finite dimensional. This theorem implies that the empty list () can be extended to a basis of $V$. In particular, $V$ has a basis.*

**Step j**

  If $w_j$ is in the span of $B$, leave $B$ unchanged. If $w_j$ is not in the
span of $B$, extend $B$ by adjoining $w_j$ to it.

After each step, $B$ is still linearly independent because otherwise the
linear dependence lemma (2.4) would give a contradiction (recall that
$(v_1, \dots, v_m)$ is linearly independent and any $w_j$ that is adjoined to $B$ is
not in the span of the previous vectors in $B$). After step $n$, the span of
$B$ includes all the $w$'s. Thus the $B$ obtained after step $n$ spans $V$ and
hence is a basis of $V$.                    ∎

  As a nice application of the theorem above, we now show that every subspace of a finite-dimensional vector space can be paired with
another subspace to form a direct sum of the whole space.

*Using the same basic*   **2.13**   **Proposition:**   *Suppose $V$ is finite dimensional and $U$ is a sub-*
*ideas but considerably*   *space of $V$. Then there is a subspace $W$ of $V$ such that $V = U \oplus W$.*
*more advanced tools,*
*this proposition can be*    PROOF:   Because $V$ is finite dimensional, so is $U$ (see 2.7). Thus
*proved without the*   there is a basis $(u_1, \dots, u_m)$ of $U$ (see 2.11). Of course $(u_1, \dots, u_m)$
*hypothesis that $V$ is*   is a linearly independent list of vectors in $V$, and thus it can be ex-
*finite dimensional.*   tended to a basis $(u_1, \dots, u_m, w_1, \dots, w_n)$ of $V$ (see 2.12). Let $W =$
$\mathrm{span}(w_1, \dots, w_n)$.

  To prove that $V = U \oplus W$, we need to show that

$$V = U + W \quad \text{and} \quad U \cap W = \{0\};$$

see 1.9. To prove the first equation, suppose that $v \in V$. Then,
because the list $(u_1, \dots, u_m, w_1, \dots, w_n)$ spans $V$, there exist scalars
$a_1, \dots, a_m, b_1, \dots, b_n \in \mathbf{F}$ such that

$$v = \underbrace{a_1 u_1 + \cdots + a_m u_m}_{u} + \underbrace{b_1 w_1 + \cdots + b_n w_n}_{w}.$$

In other words, we have $v = u + w$, where $u \in U$ and $w \in W$ are defined
as above. Thus $v \in U + W$, completing the proof that $V = U + W$.

  To show that $U \cap W = \{0\}$, suppose $v \in U \cap W$. Then there exist
scalars $a_1, \dots, a_m, b_1, \dots, b_n \in \mathbf{F}$ such that

$$v = a_1 u_1 + \cdots + a_m u_m = b_1 w_1 + \cdots + b_n w_n.$$

Thus

$$a_1u_1 + \cdots + a_mu_m - b_1w_1 - \cdots - b_nw_n = 0.$$

Because $(u_1, \ldots, u_m, w_1, \ldots, w_n)$ is linearly independent, this implies that $a_1 = \cdots = a_m = b_1 = \cdots = b_n = 0$. Thus $v = 0$, completing the proof that $U \cap W = \{0\}$. ∎

# Dimension

Though we have been discussing finite-dimensional vector spaces, we have not yet defined the dimension of such an object. How should dimension be defined? A reasonable definition should force the dimension of $\mathbf{F}^n$ to equal $n$. Notice that the basis

$$((1, 0, \ldots, 0), (0, 1, 0, \ldots, 0), \ldots, (0, \ldots, 0, 1))$$

has length $n$. Thus we are tempted to define the dimension as the length of a basis. However, a finite-dimensional vector space in general has many different bases, and our attempted definition makes sense only if all bases in a given vector space have the same length. Fortunately that turns out to be the case, as we now show.

**2.14   Theorem:** *Any two bases of a finite-dimensional vector space have the same length.*

PROOF: Suppose $V$ is finite dimensional. Let $B_1$ and $B_2$ be any two bases of $V$. Then $B_1$ is linearly independent in $V$ and $B_2$ spans $V$, so the length of $B_1$ is at most the length of $B_2$ (by 2.6). Interchanging the roles of $B_1$ and $B_2$, we also see that the length of $B_2$ is at most the length of $B_1$. Thus the length of $B_1$ must equal the length of $B_2$, as desired. ∎

Now that we know that any two bases of a finite-dimensional vector space have the same length, we can formally define the dimension of such spaces. The ***dimension*** of a finite-dimensional vector space is defined to be the length of any basis of the vector space. The dimension of $V$ (if $V$ is finite dimensional) is denoted by $\dim V$. As examples, note that $\dim \mathbf{F}^n = n$ and $\dim \mathcal{P}_m(\mathbf{F}) = m + 1$.

Every subspace of a finite-dimensional vector space is finite dimensional (by 2.7) and so has a dimension. The next result gives the expected inequality about the dimension of a subspace.

**2.15    Proposition:**    *If $V$ is finite dimensional and $U$ is a subspace of $V$, then* $\dim U \le \dim V$.

PROOF:    Suppose that $V$ is finite dimensional and $U$ is a subspace of $V$. Any basis of $U$ is a linearly independent list of vectors in $V$ and thus can be extended to a basis of $V$ (by 2.12). Hence the length of a basis of $U$ is less than or equal to the length of a basis of $V$.    ■

*The real vector space $\mathbf{R}^2$ has dimension 2; the complex vector space $\mathbf{C}$ has dimension 1. As sets, $\mathbf{R}^2$ can be identified with $\mathbf{C}$ (and addition is the same on both spaces, as is scalar multiplication by real numbers). Thus when we talk about the dimension of a vector space, the role played by the choice of $\mathbf{F}$ cannot be neglected.*

To check that a list of vectors in $V$ is a basis of $V$, we must, according to the definition, show that the list in question satisfies two properties: it must be linearly independent and it must span $V$. The next two results show that if the list in question has the right length, then we need only check that it satisfies one of the required two properties. We begin by proving that every spanning list with the right length is a basis.

**2.16    Proposition:**    *If $V$ is finite dimensional, then every spanning list of vectors in $V$ with length $\dim V$ is a basis of $V$.*

PROOF:    Suppose $\dim V = n$ and $(v_1, \ldots, v_n)$ spans $V$. The list $(v_1, \ldots, v_n)$ can be reduced to a basis of $V$ (by 2.10). However, every basis of $V$ has length $n$, so in this case the reduction must be the trivial one, meaning that no elements are deleted from $(v_1, \ldots, v_n)$. In other words, $(v_1, \ldots, v_n)$ is a basis of $V$, as desired.    ■

Now we prove that linear independence alone is enough to ensure that a list with the right length is a basis.

**2.17    Proposition:**    *If $V$ is finite dimensional, then every linearly independent list of vectors in $V$ with length $\dim V$ is a basis of $V$.*

PROOF:    Suppose $\dim V = n$ and $(v_1, \ldots, v_n)$ is linearly independent in $V$. The list $(v_1, \ldots, v_n)$ can be extended to a basis of $V$ (by 2.12). However, every basis of $V$ has length $n$, so in this case the extension must be the trivial one, meaning that no elements are adjoined to $(v_1, \ldots, v_n)$. In other words, $(v_1, \ldots, v_n)$ is a basis of $V$, as desired.    ■

As an example of how the last proposition can be applied, consider the list $((5, 7), (4, 3))$. This list of two vectors in $\mathbf{F}^2$ is obviously linearly independent (because neither vector is a scalar multiple of the other).

Because $\mathbf{F}^2$ has dimension 2, the last proposition implies that this linearly independent list of length 2 is a basis of $\mathbf{F}^2$ (we do not need to bother checking that it spans $\mathbf{F}^2$).

The next theorem gives a formula for the dimension of the sum of two subspaces of a finite-dimensional vector space.

**2.18   Theorem:**   *If $U_1$ and $U_2$ are subspaces of a finite-dimensional vector space, then*

$$\dim(U_1 + U_2) = \dim U_1 + \dim U_2 - \dim(U_1 \cap U_2).$$

*This formula for the dimension of the sum of two subspaces is analogous to a familiar counting formula: the number of elements in the union of two finite sets equals the number of elements in the first set, plus the number of elements in the second set, minus the number of elements in the intersection of the two sets.*

PROOF:   Let $(u_1, \ldots, u_m)$ be a basis of $U_1 \cap U_2$; thus $\dim(U_1 \cap U_2) = m$. Because $(u_1, \ldots, u_m)$ is a basis of $U_1 \cap U_2$, it is linearly independent in $U_1$ and hence can be extended to a basis $(u_1, \ldots, u_m, v_1, \ldots, v_j)$ of $U_1$ (by 2.12). Thus $\dim U_1 = m + j$. Also extend $(u_1, \ldots, u_m)$ to a basis $(u_1, \ldots, u_m, w_1, \ldots, w_k)$ of $U_2$; thus $\dim U_2 = m + k$.

We will show that $(u_1, \ldots, u_m, v_1, \ldots, v_j, w_1, \ldots, w_k)$ is a basis of $U_1 + U_2$. This will complete the proof because then we will have

$$
\begin{aligned}
\dim(U_1 + U_2) &= m + j + k \\
&= (m + j) + (m + k) - m \\
&= \dim U_1 + \dim U_2 - \dim(U_1 \cap U_2).
\end{aligned}
$$

Clearly $\operatorname{span}(u_1, \ldots, u_m, v_1, \ldots, v_j, w_1, \ldots, w_k)$ contains $U_1$ and $U_2$ and hence contains $U_1 + U_2$. So to show that this list is a basis of $U_1 + U_2$ we need only show that it is linearly independent. To prove this, suppose

$$a_1 u_1 + \cdots + a_m u_m + b_1 v_1 + \cdots + b_j v_j + c_1 w_1 + \cdots + c_k w_k = 0,$$

where all the $a$'s, $b$'s, and $c$'s are scalars. We need to prove that all the $a$'s, $b$'s, and $c$'s equal 0. The equation above can be rewritten as

$$c_1 w_1 + \cdots + c_k w_k = -a_1 u_1 - \cdots - a_m u_m - b_1 v_1 - \cdots - b_j v_j,$$

which shows that $c_1 w_1 + \cdots + c_k w_k \in U_1$. All the $w$'s are in $U_2$, so this implies that $c_1 w_1 + \cdots + c_k w_k \in U_1 \cap U_2$. Because $(u_1, \ldots, u_m)$ is a basis of $U_1 \cap U_2$, we can write

$$c_1 w_1 + \cdots + c_k w_k = d_1 u_1 + \cdots + d_m u_m$$

for some choice of scalars $d_1, \ldots, d_m$. But $(u_1, \ldots, u_m, w_1, \ldots, w_k)$ is linearly independent, so the last equation implies that all the $c$'s (and $d$'s) equal 0. Thus our original equation involving the $a$'s, $b$'s, and $c$'s becomes

$$a_1 u_1 + \cdots + a_m u_m + b_1 v_1 + \cdots + b_j v_j = 0.$$

This equation implies that all the $a$'s and $b$'s are 0 because the list $(u_1, \ldots, u_m, v_1, \ldots, v_j)$ is linearly independent. We now know that all the $a$'s, $b$'s, and $c$'s equal 0, as desired.                                               ∎

The next proposition shows that dimension meshes well with direct sums. This result will be useful in later chapters.

*Recall that direct sum is analogous to disjoint union. Thus 2.19 is analogous to the statement that if a finite set B is written as $A_1 \cup \cdots \cup A_m$ and the sum of the number of elements in the A's equals the number of elements in B, then the union is a disjoint union.*

**2.19  Proposition:** *Suppose $V$ is finite dimensional and $U_1, \ldots, U_m$ are subspaces of $V$ such that*

**2.20**                          $V = U_1 + \cdots + U_m$

*and*

**2.21**                          $\dim V = \dim U_1 + \cdots + \dim U_m.$

*Then $V = U_1 \oplus \cdots \oplus U_m$.*

PROOF:   Choose a basis for each $U_j$. Put these bases together in one list, forming a list that spans $V$ (by 2.20) and has length $\dim V$ (by 2.21). Thus this list is a basis of $V$ (by 2.16), and in particular it is linearly independent.

Now suppose that $u_1 \in U_1, \ldots, u_m \in U_m$ are such that

$$0 = u_1 + \cdots + u_m.$$

We can write each $u_j$ as a linear combination of the basis vectors (chosen above) of $U_j$. Substituting these linear combinations into the expression above, we have written 0 as a linear combination of the basis of $V$ constructed above. Thus all the scalars used in this linear combination must be 0. Thus each $u_j = 0$, which proves that $V = U_1 \oplus \cdots \oplus U_m$ (by 1.8).                                               ∎

# Exercises

1.  Prove that if $(v_1, \ldots, v_n)$ spans $V$, then so does the list

    $$(v_1 - v_2, v_2 - v_3, \ldots, v_{n-1} - v_n, v_n)$$

    obtained by subtracting from each vector (except the last one) the following vector.

2.  Prove that if $(v_1, \ldots, v_n)$ is linearly independent in $V$, then so is the list

    $$(v_1 - v_2, v_2 - v_3, \ldots, v_{n-1} - v_n, v_n)$$

    obtained by subtracting from each vector (except the last one) the following vector.

3.  Suppose $(v_1, \ldots, v_n)$ is linearly independent in $V$ and $w \in V$. Prove that if $(v_1 + w, \ldots, v_n + w)$ is linearly dependent, then $w \in \mathrm{span}(v_1, \ldots, v_n)$.

4.  Suppose $m$ is a positive integer. Is the set consisting of 0 and all polynomials with coefficients in $\mathbf{F}$ and with degree equal to $m$ a subspace of $\mathcal{P}(\mathbf{F})$?

5.  Prove that $\mathbf{F}^\infty$ is infinite dimensional.

6.  Prove that the real vector space consisting of all continuous real-valued functions on the interval $[0, 1]$ is infinite dimensional.

7.  Prove that $V$ is infinite dimensional if and only if there is a sequence $v_1, v_2, \ldots$ of vectors in $V$ such that $(v_1, \ldots, v_n)$ is linearly independent for every positive integer $n$.

8.  Let $U$ be the subspace of $\mathbf{R}^5$ defined by

    $$U = \{(x_1, x_2, x_3, x_4, x_5) \in \mathbf{R}^5 : x_1 = 3x_2 \text{ and } x_3 = 7x_4\}.$$

    Find a basis of $U$.

9.  Prove or disprove: there exists a basis $(p_0, p_1, p_2, p_3)$ of $\mathcal{P}_3(\mathbf{F})$ such that none of the polynomials $p_0, p_1, p_2, p_3$ has degree 2.

10. Suppose that $V$ is finite dimensional, with $\dim V = n$. Prove that there exist one-dimensional subspaces $U_1, \ldots, U_n$ of $V$ such that

    $$V = U_1 \oplus \cdots \oplus U_n.$$

LIVERPOOL
JOHN MOORES UNIVERSITY
AVRIL ROBARTS LRC
TEL. 0151 231 4022

11.  Suppose that $V$ is finite dimensional and $U$ is a subspace of $V$ such that $\dim U = \dim V$. Prove that $U = V$.

12.  Suppose that $p_0, p_1, \ldots, p_m$ are polynomials in $\mathcal{P}_m(\mathbf{F})$ such that $p_j(2) = 0$ for each $j$. Prove that $(p_0, p_1, \ldots, p_m)$ is not linearly independent in $\mathcal{P}_m(\mathbf{F})$.

13.  Suppose $U$ and $W$ are subspaces of $\mathbf{R}^8$ such that $\dim U = 3$, $\dim W = 5$, and $U + W = \mathbf{R}^8$. Prove that $U \cap W = \{0\}$.

14.  Suppose that $U$ and $W$ are both five-dimensional subspaces of $\mathbf{R}^9$. Prove that $U \cap W \neq \{0\}$.

15.  You might guess, by analogy with the formula for the number of elements in the union of three subsets of a finite set, that if $U_1, U_2, U_3$ are subspaces of a finite-dimensional vector space, then

$$\dim(U_1 + U_2 + U_3)$$
$$= \dim U_1 + \dim U_2 + \dim U_3$$
$$- \dim(U_1 \cap U_2) - \dim(U_1 \cap U_3) - \dim(U_2 \cap U_3)$$
$$+ \dim(U_1 \cap U_2 \cap U_3).$$

Prove this or give a counterexample.

16.  Prove that if $V$ is finite dimensional and $U_1, \ldots, U_m$ are subspaces of $V$, then

$$\dim(U_1 + \cdots + U_m) \le \dim U_1 + \cdots + \dim U_m.$$

17.  Suppose $V$ is finite dimensional. Prove that if $U_1, \ldots, U_m$ are subspaces of $V$ such that $V = U_1 \oplus \cdots \oplus U_m$, then

$$\dim V = \dim U_1 + \cdots + \dim U_m.$$

*This exercise deepens the analogy between direct sums of subspaces and disjoint unions of subsets. Specifically, compare this exercise to the following obvious statement: if a finite set is written as a disjoint union of subsets, then the number of elements in the set equals the sum of the number of elements in the disjoint subsets.*

# CHAPTER 3

# *Linear Maps*

So far our attention has focused on vector spaces. No one gets excited about vector spaces. The interesting part of linear algebra is the subject to which we now turn—linear maps.

Let's review our standing assumptions:

> Recall that **F** denotes **R** or **C**.
> Recall also that $V$ is a vector space over **F**.

In this chapter we will frequently need another vector space in addition to $V$. We will call this additional vector space $W$:

> Let's agree that for the rest of this chapter
> $W$ will denote a vector space over **F**.

# Definitions and Examples

*Some mathematicians use the term **linear transformation**, which means the same as linear map.*

A **linear map** from $V$ to $W$ is a function $T : V \to W$ with the following properties:

**additivity**

$\quad T(u + v) = Tu + Tv$ for all $u, v \in V$;

**homogeneity**

$\quad T(av) = a(Tv)$ for all $a \in \mathbf{F}$ and all $v \in V$.

Note that for linear maps we often use the notation $Tv$ as well as the more standard functional notation $T(v)$.

The set of all linear maps from $V$ to $W$ is denoted $\mathcal{L}(V, W)$. Let's look at some examples of linear maps. Make sure you verify that each of the functions defined below is indeed a linear map:

**zero**

In addition to its other uses, we let the symbol 0 denote the function that takes each element of some vector space to the additive identity of another vector space. To be specific, $0 \in \mathcal{L}(V, W)$ is defined by

$$0v = 0.$$

Note that the 0 on the left side of the equation above is a function from $V$ to $W$, whereas the 0 on the right side is the additive identity in $W$. As usual, the context should allow you to distinguish between the many uses of the symbol 0.

**identity**

The **identity map**, denoted $I$, is the function on some vector space that takes each element to itself. To be specific, $I \in \mathcal{L}(V, V)$ is defined by

$$Iv = v.$$

**differentiation**

Define $T \in \mathcal{L}(\mathcal{P}(\mathbf{R}), \mathcal{P}(\mathbf{R}))$ by

$$Tp = p'.$$

The assertion that this function is a linear map is another way of stating a basic result about differentiation: $(f + g)' = f' + g'$ and $(af)' = af'$ whenever $f, g$ are differentiable and $a$ is a constant.

## integration

Define $T \in \mathcal{L}(\mathcal{P}(\mathbf{R}), \mathbf{R})$ by

$$Tp = \int_0^1 p(x)\, dx.$$

The assertion that this function is linear is another way of stating a basic result about integration: the integral of the sum of two functions equals the sum of the integrals, and the integral of a constant times a function equals the constant times the integral of the function.

## multiplication by $x^2$

Define $T \in \mathcal{L}(\mathcal{P}(\mathbf{R}), \mathcal{P}(\mathbf{R}))$ by

$$(Tp)(x) = x^2 p(x)$$

for $x \in \mathbf{R}$.

## backward shift

Recall that $\mathbf{F}^\infty$ denotes the vector space of all sequences of elements of $\mathbf{F}$. Define $T \in \mathcal{L}(\mathbf{F}^\infty, \mathbf{F}^\infty)$ by

$$T(x_1, x_2, x_3, \dots) = (x_2, x_3, \dots).$$

## from $\mathbf{F}^n$ to $\mathbf{F}^m$

Define $T \in \mathcal{L}(\mathbf{R}^3, \mathbf{R}^2)$ by

$$T(x, y, z) = (2x - y + 3z, 7x + 5y - 6z).$$

More generally, let $m$ and $n$ be positive integers, let $a_{j,k} \in \mathbf{F}$ for $j = 1, \dots, m$ and $k = 1, \dots, n$, and define $T \in \mathcal{L}(\mathbf{F}^n, \mathbf{F}^m)$ by

$$T(x_1, \dots, x_n) = (a_{1,1}x_1 + \cdots + a_{1,n}x_n, \dots, a_{m,1}x_1 + \cdots + a_{m,n}x_n).$$

Later we will see that every linear map from $\mathbf{F}^n$ to $\mathbf{F}^m$ is of this form.

Suppose $(v_1, \dots, v_n)$ is a basis of $V$ and $T \colon V \to W$ is linear. If $v \in V$, then we can write $v$ in the form

$$v = a_1 v_1 + \cdots + a_n v_n.$$

The linearity of $T$ implies that

*Though linear maps are pervasive throughout mathematics, they are not as ubiquitous as imagined by some confused students who seem to think that $\cos$ is a linear map from $\mathbf{R}$ to $\mathbf{R}$ when they write "identities" such as $\cos 2x = 2 \cos x$ and $\cos(x + y) = \cos x + \cos y$.*

$$Tv = a_1 Tv_1 + \cdots + a_n Tv_n.$$

In particular, the values of $Tv_1, \ldots, Tv_n$ determine the values of $T$ on arbitrary vectors in $V$.

Linear maps can be constructed that take on arbitrary values on a basis. Specifically, given a basis $(v_1, \ldots, v_n)$ of $V$ and any choice of vectors $w_1, \ldots, w_n \in W$, we can construct a linear map $T \colon V \to W$ such that $Tv_j = w_j$ for $j = 1, \ldots, n$. There is no choice of how to do this—we must define $T$ by

$$T(a_1 v_1 + \cdots + a_n v_n) = a_1 w_1 + \cdots + a_n w_n,$$

where $a_1, \ldots, a_n$ are arbitrary elements of $\mathbf{F}$. Because $(v_1, \ldots, v_n)$ is a basis of $V$, the equation above does indeed define a function $T$ from $V$ to $W$. You should verify that the function $T$ defined above is linear and that $Tv_j = w_j$ for $j = 1, \ldots, n$.

Now we will make $\mathcal{L}(V, W)$ into a vector space by defining addition and scalar multiplication on it. For $S, T \in \mathcal{L}(V, W)$, define a function $S + T \in \mathcal{L}(V, W)$ in the usual manner of adding functions:

$$(S + T)v = Sv + Tv$$

for $v \in V$. You should verify that $S + T$ is indeed a linear map from $V$ to $W$ whenever $S, T \in \mathcal{L}(V, W)$. For $a \in \mathbf{F}$ and $T \in \mathcal{L}(V, W)$, define a function $aT \in \mathcal{L}(V, W)$ in the usual manner of multiplying a function by a scalar:

$$(aT)v = a(Tv)$$

for $v \in V$. You should verify that $aT$ is indeed a linear map from $V$ to $W$ whenever $a \in \mathbf{F}$ and $T \in \mathcal{L}(V, W)$. With the operations we have just defined, $\mathcal{L}(V, W)$ becomes a vector space (as you should verify). Note that the additive identity of $\mathcal{L}(V, W)$ is the zero linear map defined earlier in this section.

Usually it makes no sense to multiply together two elements of a vector space, but for some pairs of linear maps a useful product exists. We will need a third vector space, so suppose $U$ is a vector space over $\mathbf{F}$. If $T \in \mathcal{L}(U, V)$ and $S \in \mathcal{L}(V, W)$, then we define $ST \in \mathcal{L}(U, W)$ by

$$(ST)(v) = S(Tv)$$

for $v \in U$. In other words, $ST$ is just the usual composition $S \circ T$ of two functions, but when both functions are linear, most mathematicians write $ST$ instead of $S \circ T$. You should verify that $ST$ is indeed a linear map from $U$ to $W$ whenever $T \in \mathcal{L}(U, V)$ and $S \in \mathcal{L}(V, W)$. Note that $ST$ is defined only when $T$ maps into the domain of $S$. We often call $ST$ the **product** of $S$ and $T$. You should verify that it has most of the usual properties expected of a product:

**associativity**
> $(T_1 T_2) T_3 = T_1 (T_2 T_3)$ whenever $T_1$, $T_2$, and $T_3$ are linear maps such that the products make sense (meaning that $T_3$ must map into the domain of $T_2$, and $T_2$ must map into the domain of $T_1$).

**identity**
> $TI = T$ and $IT = T$ whenever $T \in \mathcal{L}(V, W)$ (note that in the first equation $I$ is the identity map on $V$, and in the second equation $I$ is the identity map on $W$).

**distributive properties**
> $(S_1 + S_2) T = S_1 T + S_2 T$ and $S(T_1 + T_2) = S T_1 + S T_2$ whenever $T, T_1, T_2 \in \mathcal{L}(U, V)$ and $S, S_1, S_2 \in \mathcal{L}(V, W)$.

Multiplication of linear maps is not commutative. In other words, it is not necessarily true that $ST = TS$, even if both sides of the equation make sense. For example, if $T \in \mathcal{L}(\mathcal{P}(\mathbf{R}), \mathcal{P}(\mathbf{R}))$ is the differentiation map defined earlier in this section and $S \in \mathcal{L}(\mathcal{P}(\mathbf{R}), \mathcal{P}(\mathbf{R}))$ is the multiplication by $x^2$ map defined earlier in this section, then

$$((ST)p)(x) = x^2 p'(x) \quad \text{but} \quad ((TS)p)(x) = x^2 p'(x) + 2x p(x).$$

In other words, multiplying by $x^2$ and then differentiating is not the same as differentiating and then multiplying by $x^2$.

## Null Spaces and Ranges

For $T \in \mathcal{L}(V, W)$, the **null space** of $T$, denoted null $T$, is the subset of $V$ consisting of those vectors that $T$ maps to 0:

$$\text{null } T = \{v \in V : Tv = 0\}.$$

*Some mathematicians use the term **kernel** instead of null space.*

Let's look at a few examples from the previous section. In the differentiation example, we defined $T \in \mathcal{L}(\mathcal{P}(\mathbf{R}), \mathcal{P}(\mathbf{R}))$ by $Tp = p'$. The

LIVERPOOL JOHN MOORES UNIVERSITY
LEARNING SERVICES

only functions whose derivative equals the zero function are the constant functions, so in this case the null space of $T$ equals the set of constant functions.

In the multiplication by $x^2$ example, we defined $T \in \mathcal{L}(\mathcal{P}(\mathbf{R}), \mathcal{P}(\mathbf{R}))$ by $(Tp)(x) = x^2 p(x)$. The only polynomial $p$ such that $x^2 p(x) = 0$ for all $x \in \mathbf{R}$ is the 0 polynomial. Thus in this case we have

$$\text{null } T = \{0\}.$$

In the backward shift example, we defined $T \in \mathcal{L}(\mathbf{F}^\infty, \mathbf{F}^\infty)$ by

$$T(x_1, x_2, x_3, \dots) = (x_2, x_3, \dots).$$

Clearly $T(x_1, x_2, x_3, \dots)$ equals 0 if and only if $x_2, x_3, \dots$ are all 0. Thus in this case we have

$$\text{null } T = \{(a, 0, 0, \dots) : a \in \mathbf{F}\}.$$

The next proposition shows that the null space of any linear map is a subspace of the domain. In particular, 0 is in the null space of every linear map.

**3.1    Proposition:**  *If $T \in \mathcal{L}(V, W)$, then* null $T$ *is a subspace of $V$.*

PROOF:  Suppose $T \in \mathcal{L}(V, W)$. By additivity, we have

$$T(0) = T(0 + 0) = T(0) + T(0),$$

which implies that $T(0) = 0$. Thus $0 \in \text{null } T$.
  If $u, v \in \text{null } T$, then

$$T(u + v) = Tu + Tv = 0 + 0 = 0,$$

and hence $u + v \in \text{null } T$. Thus null $T$ is closed under addition.
  If $u \in \text{null } T$ and $a \in \mathbf{F}$, then

$$T(au) = aTu = a0 = 0,$$

and hence $au \in \text{null } T$. Thus null $T$ is closed under scalar multiplication.

We have shown that null $T$ contains 0 and is closed under addition and scalar multiplication. Thus null $T$ is a subspace of $V$. ∎

A linear map $T: V \rightarrow W$ is called **injective** if whenever $u, v \in V$ and $Tu = Tv$, we have $u = v$. The next proposition says that we can check whether a linear map is injective by checking whether 0 is the only vector that gets mapped to 0. As a simple application of this proposition, we see that of the three linear maps whose null spaces we computed earlier in this section (differentiation, multiplication by $x^2$, and backward shift), only multiplication by $x^2$ is injective.

*Many mathematicians use the term **one-to-one**, which means the same as injective.*

**3.2     Proposition:** *Let $T \in \mathcal{L}(V, W)$. Then $T$ is injective if and only if* null $T = \{0\}$.

PROOF:     First suppose that $T$ is injective. We want to prove that null $T = \{0\}$. We already know that $\{0\} \subset$ null $T$ (by 3.1). To prove the inclusion in the other direction, suppose $v \in$ null $T$. Then

$$T(v) = 0 = T(0).$$

Because $T$ is injective, the equation above implies that $v = 0$. Thus null $T = \{0\}$, as desired.

To prove the implication in the other direction, now suppose that null $T = \{0\}$. We want to prove that $T$ is injective. To do this, suppose $u, v \in V$ and $Tu = Tv$. Then

$$0 = Tu - Tv = T(u - v).$$

Thus $u - v$ is in null $T$, which equals $\{0\}$. Hence $u - v = 0$, which implies that $u = v$. Hence $T$ is injective, as desired.     ■

For $T \in \mathcal{L}(V, W)$, the **range** of $T$, denoted range $T$, is the subset of $W$ consisting of those vectors that are of the form $Tv$ for some $v \in V$:

$$\text{range } T = \{Tv : v \in V\}.$$

*Some mathematicians use the word **image**, which means the same as range.*

For example, if $T \in \mathcal{L}(\mathcal{P}(\mathbf{R}), \mathcal{P}(\mathbf{R}))$ is the differentiation map defined by $Tp = p'$, then range $T = \mathcal{P}(\mathbf{R})$ because for every polynomial $q \in \mathcal{P}(\mathbf{R})$ there exists a polynomial $p \in \mathcal{P}(\mathbf{R})$ such that $p' = q$.

As another example, if $T \in \mathcal{L}(\mathcal{P}(\mathbf{R}), \mathcal{P}(\mathbf{R}))$ is the linear map of multiplication by $x^2$ defined by $(Tp)(x) = x^2 p(x)$, then the range of $T$ is the set of polynomials of the form $a_2 x^2 + \cdots + a_m x^m$, where $a_2, \ldots, a_m \in \mathbf{R}$.

The next proposition shows that the range of any linear map is a subspace of the target space.

**3.3    Proposition:**  *If $T \in \mathcal{L}(V, W)$, then* range $T$ *is a subspace of $W$.*

PROOF:   Suppose $T \in \mathcal{L}(V, W)$. Then $T(0) = 0$ (by 3.1), which implies that $0 \in$ range $T$.

If $w_1, w_2 \in$ range $T$, then there exist $v_1, v_2 \in V$ such that $Tv_1 = w_1$ and $Tv_2 = w_2$. Thus

$$T(v_1 + v_2) = Tv_1 + Tv_2 = w_1 + w_2,$$

and hence $w_1 + w_2 \in$ range $T$. Thus range $T$ is closed under addition.

If $w \in$ range $T$ and $a \in \mathbf{F}$, then there exists $v \in V$ such that $Tv = w$. Thus

$$T(av) = aTv = aw,$$

and hence $aw \in$ range $T$. Thus range $T$ is closed under scalar multiplication.

We have shown that range $T$ contains 0 and is closed under addition and scalar multiplication. Thus range $T$ is a subspace of $W$.   ∎

*Many mathematicians use the term **onto**, which means the same as surjective.*

A linear map $T : V \to W$ is called ***surjective*** if its range equals $W$. For example, the differentiation map $T \in \mathcal{L}(\mathcal{P}(\mathbf{R}), \mathcal{P}(\mathbf{R}))$ defined by $Tp = p'$ is surjective because its range equals $\mathcal{P}(\mathbf{R})$. As another example, the linear map $T \in \mathcal{L}(\mathcal{P}(\mathbf{R}), \mathcal{P}(\mathbf{R}))$ defined by $(Tp)(x) = x^2 p(x)$ is not surjective because its range does not equal $\mathcal{P}(\mathbf{R})$. As a final example, you should verify that the backward shift $T \in \mathcal{L}(\mathbf{F}^\infty, \mathbf{F}^\infty)$ defined by

$$T(x_1, x_2, x_3, \dots) = (x_2, x_3, \dots)$$

is surjective.

Whether a linear map is surjective can depend upon what we are thinking of as the target space. For example, fix a positive integer $m$. The differentiation map $T \in \mathcal{L}(\mathcal{P}_m(\mathbf{R}), \mathcal{P}_m(\mathbf{R}))$ defined by $Tp = p'$ is not surjective because the polynomial $x^m$ is not in the range of $T$. However, the differentiation map $T \in \mathcal{L}(\mathcal{P}_m(\mathbf{R}), \mathcal{P}_{m-1}(\mathbf{R}))$ defined by $Tp = p'$ is surjective because its range equals $\mathcal{P}_{m-1}(\mathbf{R})$, which is now the target space.

The next theorem, which is the key result in this chapter, states that the dimension of the null space plus the dimension of the range of a linear map on a finite-dimensional vector space equals the dimension of the domain.

**3.4    Theorem:** *If $V$ is finite dimensional and $T \in \mathcal{L}(V, W)$, then range $T$ is a finite-dimensional subspace of $W$ and*

$$\dim V = \dim \operatorname{null} T + \dim \operatorname{range} T.$$

PROOF: Suppose that $V$ is a finite-dimensional vector space and $T \in \mathcal{L}(V, W)$. Let $(u_1, \ldots, u_m)$ be a basis of null $T$; thus $\dim \operatorname{null} T = m$. The linearly independent list $(u_1, \ldots, u_m)$ can be extended to a basis $(u_1, \ldots, u_m, w_1, \ldots, w_n)$ of $V$ (by 2.12). Thus $\dim V = m + n$, and to complete the proof, we need only show that range $T$ is finite dimensional and $\dim \operatorname{range} T = n$. We will do this by proving that $(Tw_1, \ldots, Tw_n)$ is a basis of range $T$.

Let $v \in V$. Because $(u_1, \ldots, u_m, w_1, \ldots, w_n)$ spans $V$, we can write

$$v = a_1 u_1 + \cdots + a_m u_m + b_1 w_1 + \cdots + b_n w_n,$$

where the $a$'s and $b$'s are in $\mathbf{F}$. Applying $T$ to both sides of this equation, we get

$$Tv = b_1 Tw_1 + \cdots + b_n Tw_n,$$

where the terms of the form $Tu_j$ disappeared because each $u_j \in \operatorname{null} T$. The last equation implies that $(Tw_1, \ldots, Tw_n)$ spans range $T$. In particular, range $T$ is finite dimensional.

To show that $(Tw_1, \ldots, Tw_n)$ is linearly independent, suppose that $c_1, \ldots, c_n \in \mathbf{F}$ and

$$c_1 Tw_1 + \cdots + c_n Tw_n = 0.$$

Then

$$T(c_1 w_1 + \cdots + c_n w_n) = 0,$$

and hence

$$c_1 w_1 + \cdots + c_n w_n \in \operatorname{null} T.$$

Because $(u_1, \ldots, u_m)$ spans null $T$, we can write

$$c_1 w_1 + \cdots + c_n w_n = d_1 u_1 + \cdots + d_m u_m,$$

where the $d$'s are in $\mathbf{F}$. This equation implies that all the $c$'s (and $d$'s) are 0 (because $(u_1, \ldots, u_m, w_1, \ldots, w_n)$ is linearly independent). Thus $(Tw_1, \ldots, Tw_n)$ is linearly independent and hence is a basis for range $T$, as desired. ∎

Now we can show that no linear map from a finite-dimensional vector space to a "smaller" vector space can be injective, where "smaller" is measured by dimension.

**3.5**   **Corollary:**  *If $V$ and $W$ are finite-dimensional vector spaces such that $\dim V > \dim W$, then no linear map from $V$ to $W$ is injective.*

PROOF:  Suppose $V$ and $W$ are finite-dimensional vector spaces such that $\dim V > \dim W$. Let $T \in \mathcal{L}(V, W)$. Then

$$\begin{aligned}
\dim \operatorname{null} T &= \dim V - \dim \operatorname{range} T \\
&\geq \dim V - \dim W \\
&> 0,
\end{aligned}$$

where the equality above comes from 3.4. We have just shown that $\dim \operatorname{null} T > 0$. This means that $\operatorname{null} T$ must contain vectors other than 0. Thus $T$ is not injective (by 3.2).    ∎

The next corollary, which is in some sense dual to the previous corollary, shows that no linear map from a finite-dimensional vector space to a "bigger" vector space can be surjective, where "bigger" is measured by dimension.

**3.6**   **Corollary:**  *If $V$ and $W$ are finite-dimensional vector spaces such that $\dim V < \dim W$, then no linear map from $V$ to $W$ is surjective.*

PROOF:  Suppose $V$ and $W$ are finite-dimensional vector spaces such that $\dim V < \dim W$. Let $T \in \mathcal{L}(V, W)$. Then

$$\begin{aligned}
\dim \operatorname{range} T &= \dim V - \dim \operatorname{null} T \\
&\leq \dim V \\
&< \dim W,
\end{aligned}$$

where the equality above comes from 3.4. We have just shown that $\dim \operatorname{range} T < \dim W$. This means that $\operatorname{range} T$ cannot equal $W$. Thus $T$ is not surjective.    ∎

The last two corollaries have important consequences in the theory of linear equations. To see this, fix positive integers $m$ and $n$, and let $a_{j,k} \in \mathbf{F}$ for $j = 1, \ldots, m$ and $k = 1, \ldots, n$. Define $T\colon \mathbf{F}^n \to \mathbf{F}^m$ by

$$T(x_1,\ldots,x_n) = (\sum_{k=1}^{n} a_{1,k}x_k,\ldots,\sum_{k=1}^{n} a_{m,k}x_k).$$

Now consider the equation $Tx = 0$ (where $x \in \mathbf{F}^n$ and the 0 here is the additive identity in $\mathbf{F}^m$, namely, the list of length $m$ consisting of all 0's). Letting $x = (x_1,\ldots,x_n)$, we can rewrite the equation $Tx = 0$ as a system of homogeneous equations:

$$\sum_{k=1}^{n} a_{1,k}x_k = 0$$

$$\vdots$$

$$\sum_{k=1}^{n} a_{m,k}x_k = 0.$$

*Homogeneous, in this context, means that the constant term on the right side of each equation equals 0.*

We think of the $a$'s as known; we are interested in solutions for the variables $x_1,\ldots,x_n$. Thus we have $m$ equations and $n$ variables. Obviously $x_1 = \cdots = x_n = 0$ is a solution; the key question here is whether any other solutions exist. In other words, we want to know if null $T$ is strictly bigger than $\{0\}$. This happens precisely when $T$ is not injective (by 3.2). From 3.5 we see that $T$ is not injective if $n > m$. Conclusion: a homogeneous system of linear equations in which there are more variables than equations must have nonzero solutions.

With $T$ as in the previous paragraph, now consider the equation $Tx = c$, where $c = (c_1,\ldots,c_m) \in \mathbf{F}^m$. We can rewrite the equation $Tx = c$ as a system of inhomogeneous equations:

$$\sum_{k=1}^{n} a_{1,k}x_k = c_1$$

$$\vdots$$

$$\sum_{k=1}^{n} a_{m,k}x_k = c_m.$$

*These results about homogeneous systems with more variables than equations and inhomogeneous systems with more equations than variables are often proved using Gaussian elimination. The abstract approach taken here leads to cleaner proofs.*

As before, we think of the $a$'s as known. The key question here is whether for every choice of the constant terms $c_1,\ldots,c_m \in \mathbf{F}$, there exists at least one solution for the variables $x_1,\ldots,x_n$. In other words, we want to know whether range $T$ equals $\mathbf{F}^m$. From 3.6 we see that $T$ is not surjective if $n < m$. Conclusion: an inhomogeneous system of linear equations in which there are more equations than variables has no solution for some choice of the constant terms.

# The Matrix of a Linear Map

We have seen that if $(v_1, \ldots, v_n)$ is a basis of $V$ and $T: V \to W$ is linear, then the values of $Tv_1, \ldots, Tv_n$ determine the values of $T$ on arbitrary vectors in $V$. In this section we will see how matrices are used as an efficient method of recording the values of the $Tv_j$'s in terms of a basis of $W$.

Let $m$ and $n$ denote positive integers. An $m$-by-$n$ **matrix** is a rectangular array with $m$ rows and $n$ columns that looks like this:

**3.7**
$$\begin{bmatrix} a_{1,1} & \cdots & a_{1,n} \\ \vdots & & \vdots \\ a_{m,1} & \cdots & a_{m,n} \end{bmatrix}.$$

Note that the first index refers to the row number and the second index refers to the column number. Thus $a_{3,2}$ refers to the entry in the third row, second column of the matrix above. We will usually consider matrices whose entries are elements of $\mathbf{F}$.

Let $T \in \mathcal{L}(V, W)$. Suppose that $(v_1, \ldots, v_n)$ is a basis of $V$ and $(w_1, \ldots, w_m)$ is a basis of $W$. For each $k = 1, \ldots, n$, we can write $Tv_k$ uniquely as a linear combination of the $w$'s:

**3.8**
$$Tv_k = a_{1,k}w_1 + \cdots + a_{m,k}w_m,$$

where $a_{j,k} \in \mathbf{F}$ for $j = 1, \ldots, m$. The scalars $a_{j,k}$ completely determine the linear map $T$ because a linear map is determined by its values on a basis. The $m$-by-$n$ matrix 3.7 formed by the $a$'s is called the **matrix** of $T$ with respect to the bases $(v_1, \ldots, v_n)$ and $(w_1, \ldots, w_m)$; we denote it by

$$\mathcal{M}(T, (v_1, \ldots, v_n), (w_1, \ldots, w_m)).$$

If the bases $(v_1, \ldots, v_n)$ and $(w_1, \ldots, w_m)$ are clear from the context (for example, if only one set of bases is in sight), we write just $\mathcal{M}(T)$ instead of $\mathcal{M}(T, (v_1, \ldots, v_n), (w_1, \ldots, w_m))$.

As an aid to remembering how $\mathcal{M}(T)$ is constructed from $T$, you might write the basis vectors $v_1, \ldots, v_n$ for the domain across the top and the basis vectors $w_1, \ldots, w_m$ for the target space along the left, as follows:

$$
\begin{array}{c}
\begin{array}{ccccc} v_1 & \cdots & v_k & \cdots & v_n \end{array} \\
\begin{array}{c} w_1 \\ \vdots \\ w_m \end{array}
\left[ \begin{array}{ccc} & a_{1,k} & \\ & \vdots & \\ & a_{m,k} & \end{array} \right]
\end{array}
$$

Note that in the matrix above only the $k^{\text{th}}$ column is displayed (and thus the second index of each displayed $a$ is $k$). The $k^{\text{th}}$ column of $\mathcal{M}(T)$ consists of the scalars needed to write $Tv_k$ as a linear combination of the $w$'s. Thus the picture above should remind you that $Tv_k$ is retrieved from the matrix $\mathcal{M}(T)$ by multiplying each entry in the $k^{\text{th}}$ column by the corresponding $w$ from the left column, and then adding up the resulting vectors.

*With respect to any choice of bases, the matrix of the 0 linear map (the linear map that takes every vector to 0) consists of all 0's.*

If $T$ is a linear map from $\mathbf{F}^n$ to $\mathbf{F}^m$, then unless stated otherwise you should assume that the bases in question are the standard ones (where the $k^{\text{th}}$ basis vector is 1 in the $k^{\text{th}}$ slot and 0 in all the other slots). If you think of elements of $\mathbf{F}^m$ as columns of $m$ numbers, then you can think of the $k^{\text{th}}$ column of $\mathcal{M}(T)$ as $T$ applied to the $k^{\text{th}}$ basis vector. For example, if $T \in \mathcal{L}(\mathbf{F}^2, \mathbf{F}^3)$ is defined by

$$ T(x, y) = (x + 3y, 2x + 5y, 7x + 9y), $$

then $T(1, 0) = (1, 2, 7)$ and $T(0, 1) = (3, 5, 9)$, so the matrix of $T$ (with respect to the standard bases) is the 3-by-2 matrix

$$ \left[ \begin{array}{cc} 1 & 3 \\ 2 & 5 \\ 7 & 9 \end{array} \right]. $$

Suppose we have bases $(v_1, \dots, v_n)$ of $V$ and $(w_1, \dots, w_m)$ of $W$. Thus for each linear map from $V$ to $W$, we can talk about its matrix (with respect to these bases, of course). Is the matrix of the sum of two linear maps equal to the sum of the matrices of the two maps?

Right now this question does not make sense because, though we have defined the sum of two linear maps, we have not defined the sum of two matrices. Fortunately the obvious definition of the sum of two matrices has the right properties. Specifically, we define addition of matrices of the same size by adding corresponding entries in the matrices:

$$\begin{bmatrix} a_{1,1} & \cdots & a_{1,n} \\ \vdots & & \vdots \\ a_{m,1} & \cdots & a_{m,n} \end{bmatrix} + \begin{bmatrix} b_{1,1} & \cdots & b_{1,n} \\ \vdots & & \vdots \\ b_{m,1} & \cdots & b_{m,n} \end{bmatrix}$$
$$= \begin{bmatrix} a_{1,1} + b_{1,1} & \cdots & a_{1,n} + b_{1,n} \\ \vdots & & \vdots \\ a_{m,1} + b_{m,1} & \cdots & a_{m,n} + b_{m,n} \end{bmatrix}.$$

You should verify that with this definition of matrix addition,

**3.9**                       $$\mathcal{M}(T + S) = \mathcal{M}(T) + \mathcal{M}(S)$$

whenever $T, S \in \mathcal{L}(V, W)$.

Still assuming that we have some bases in mind, is the matrix of a scalar times a linear map equal to the scalar times the matrix of the linear map? Again the question does not make sense because we have not defined scalar multiplication on matrices. Fortunately the obvious definition again has the right properties. Specifically, we define the product of a scalar and a matrix by multiplying each entry in the matrix by the scalar:

$$c \begin{bmatrix} a_{1,1} & \cdots & a_{1,n} \\ \vdots & & \vdots \\ a_{m,1} & \cdots & a_{m,n} \end{bmatrix} = \begin{bmatrix} ca_{1,1} & \cdots & ca_{1,n} \\ \vdots & & \vdots \\ ca_{m,1} & \cdots & ca_{m,n} \end{bmatrix}.$$

You should verify that with this definition of scalar multiplication on matrices,

**3.10**                          $$\mathcal{M}(cT) = c\mathcal{M}(T)$$

whenever $c \in \mathbf{F}$ and $T \in \mathcal{L}(V, W)$.

Because addition and scalar multiplication have now been defined for matrices, you should not be surprised that a vector space is about to appear. We need only a bit of notation so that this new vector space has a name. The set of all $m$-by-$n$ matrices with entries in $\mathbf{F}$ is denoted by $\text{Mat}(m, n, \mathbf{F})$. You should verify that with addition and scalar multiplication defined as above, $\text{Mat}(m, n, \mathbf{F})$ is a vector space. Note that the additive identity in $\text{Mat}(m, n, \mathbf{F})$ is the $m$-by-$n$ matrix all of whose entries equal 0.

Suppose $(v_1, \ldots, v_n)$ is a basis of $V$ and $(w_1, \ldots, w_m)$ is a basis of $W$. Suppose also that we have another vector space $U$ and that $(u_1, \ldots, u_p)$

is a basis of $U$. Consider linear maps $S: U \to V$ and $T: V \to W$. The composition $TS$ is a linear map from $U$ to $W$. How can $\mathcal{M}(TS)$ be computed from $\mathcal{M}(T)$ and $\mathcal{M}(S)$? The nicest solution to this question would be to have the following pretty relationship:

**3.11**                               $\mathcal{M}(TS) = \mathcal{M}(T)\mathcal{M}(S)$.

So far, however, the right side of this equation does not make sense because we have not yet defined the product of two matrices. We will choose a definition of matrix multiplication that forces the equation above to hold. Let's see how to do this.

Let

$$\mathcal{M}(T) = \begin{bmatrix} a_{1,1} & \cdots & a_{1,n} \\ \vdots & & \vdots \\ a_{m,1} & \cdots & a_{m,n} \end{bmatrix} \quad \text{and} \quad \mathcal{M}(S) = \begin{bmatrix} b_{1,1} & \cdots & b_{1,p} \\ \vdots & & \vdots \\ b_{n,1} & \cdots & b_{n,p} \end{bmatrix}.$$

For $k \in \{1, \ldots, p\}$, we have

$$\begin{aligned} TSu_k &= T(\sum_{r=1}^{n} b_{r,k}v_r) \\ &= \sum_{r=1}^{n} b_{r,k}Tv_r \\ &= \sum_{r=1}^{n} b_{r,k} \sum_{j=1}^{m} a_{j,r}w_j \\ &= \sum_{j=1}^{m} (\sum_{r=1}^{n} a_{j,r}b_{r,k})w_j. \end{aligned}$$

Thus $\mathcal{M}(TS)$ is the $m$-by-$p$ matrix whose entry in row $j$, column $k$ equals $\sum_{r=1}^{n} a_{j,r}b_{r,k}$.

Now it's clear how to define matrix multiplication so that 3.11 holds. Namely, if $A$ is an $m$-by-$n$ matrix with entries $a_{j,k}$ and $B$ is an $n$-by-$p$ matrix with entries $b_{j,k}$, then $AB$ is defined to be the $m$-by-$p$ matrix whose entry in row $j$, column $k$, equals

$$\sum_{r=1}^{n} a_{j,r}b_{r,k}.$$

*You probably learned this definition of matrix multiplication in an earlier course, although you may not have seen this motivation for it.*

In other words, the entry in row $j$, column $k$, of $AB$ is computed by taking row $j$ of $A$ and column $k$ of $B$, multiplying together corresponding entries, and then summing. Note that we define the product of two matrices only when the number of columns of the first matrix equals the number of rows of the second matrix.

*You should find an example to show that matrix multiplication is not commutative. In other words, AB is not necessarily equal to BA, even when both are defined.*

As an example of matrix multiplication, here we multiply together a 3-by-2 matrix and a 2-by-4 matrix, obtaining a 3-by-4 matrix:

$$\begin{bmatrix} 1 & 2 \\ 3 & 4 \\ 5 & 6 \end{bmatrix} \begin{bmatrix} 6 & 5 & 4 & 3 \\ 2 & 1 & 0 & -1 \end{bmatrix} = \begin{bmatrix} 10 & 7 & 4 & 1 \\ 26 & 19 & 12 & 5 \\ 42 & 31 & 20 & 9 \end{bmatrix}.$$

Suppose $(v_1, \ldots, v_n)$ is a basis of $V$. If $v \in V$, then there exist unique scalars $b_1, \ldots, b_n$ such that

**3.12** $$v = b_1 v_1 + \cdots + b_n v_n.$$

The **matrix** of $v$, denoted $\mathcal{M}(v)$, is the $n$-by-1 matrix defined by

**3.13** $$\mathcal{M}(v) = \begin{bmatrix} b_1 \\ \vdots \\ b_n \end{bmatrix}.$$

Usually the basis is obvious from the context, but when the basis needs to be displayed explicitly use the notation $\mathcal{M}(v, (v_1, \ldots, v_n))$ instead of $\mathcal{M}(v)$.

For example, the matrix of a vector $x \in \mathbf{F}^n$ with respect to the standard basis is obtained by writing the coordinates of $x$ as the entries in an $n$-by-1 matrix. In other words, if $x = (x_1, \ldots, x_n) \in \mathbf{F}^n$, then

$$\mathcal{M}(x) = \begin{bmatrix} x_1 \\ \vdots \\ x_n \end{bmatrix}.$$

The next proposition shows how the notions of the matrix of a linear map, the matrix of a vector, and matrix multiplication fit together. In this proposition $\mathcal{M}(Tv)$ is the matrix of the vector $Tv$ with respect to the basis $(w_1, \ldots, w_m)$ and $\mathcal{M}(v)$ is the matrix of the vector $v$ with respect to the basis $(v_1, \ldots, v_n)$, whereas $\mathcal{M}(T)$ is the matrix of the linear map $T$ with respect to the bases $(v_1, \ldots, v_n)$ and $(w_1, \ldots, w_m)$.

**3.14  Proposition:** *Suppose $T \in \mathcal{L}(V, W)$ and $(v_1, \ldots, v_n)$ is a basis of $V$ and $(w_1, \ldots, w_m)$ is a basis of $W$. Then*

$$\mathcal{M}(Tv) = \mathcal{M}(T)\mathcal{M}(v)$$

*for every $v \in V$.*

PROOF:  Let

**3.15**                    $$\mathcal{M}(T) = \begin{bmatrix} a_{1,1} & \cdots & a_{1,n} \\ \vdots & & \vdots \\ a_{m,1} & \cdots & a_{m,n} \end{bmatrix}.$$

This means, we recall, that

**3.16**                    $$T v_k = \sum_{j=1}^{m} a_{j,k} w_j$$

for each $k$. Let $v$ be an arbitrary vector in $V$, which we can write in the form 3.12. Thus $\mathcal{M}(v)$ is given by 3.13. Now

$$\begin{aligned} Tv &= b_1 T v_1 + \cdots + b_n T v_n \\ &= b_1 \sum_{j=1}^{m} a_{j,1} w_j + \cdots + b_n \sum_{j=1}^{m} a_{j,n} w_j \\ &= \sum_{j=1}^{m} (a_{j,1} b_1 + \cdots + a_{j,n} b_n) w_j, \end{aligned}$$

where the first equality comes from 3.12 and the second equality comes from 3.16. The last equation shows that $\mathcal{M}(Tv)$, the $m$-by-1 matrix of the vector $Tv$ with respect to the basis $(w_1, \ldots, w_m)$, is given by the equation

$$\mathcal{M}(Tv) = \begin{bmatrix} a_{1,1} b_1 + \cdots + a_{1,n} b_n \\ \vdots \\ a_{m,1} b_1 + \cdots + a_{m,n} b_n \end{bmatrix}.$$

This formula, along with the formulas 3.15 and 3.13 and the definition of matrix multiplication, shows that $\mathcal{M}(Tv) = \mathcal{M}(T)\mathcal{M}(v)$.            ∎

## Invertibility

A linear map $T \in \mathcal{L}(V, W)$ is called **invertible** if there exists a linear map $S \in \mathcal{L}(W, V)$ such that $ST$ equals the identity map on $V$ and $TS$ equals the identity map on $W$. A linear map $S \in \mathcal{L}(W, V)$ satisfying $ST = I$ and $TS = I$ is called an **inverse** of $T$ (note that the first $I$ is the identity map on $V$ and the second $I$ is the identity map on $W$).

If $S$ and $S'$ are inverses of $T$, then

$$S = SI = S(TS') = (ST)S' = IS' = S',$$

so $S = S'$. In other words, if $T$ is invertible, then it has a unique inverse, which we denote by $T^{-1}$. Rephrasing all this once more, if $T \in \mathcal{L}(V, W)$ is invertible, then $T^{-1}$ is the unique element of $\mathcal{L}(W, V)$ such that $T^{-1}T = I$ and $TT^{-1} = I$. The following proposition characterizes the invertible linear maps.

**3.17    Proposition:** *A linear map is invertible if and only if it is injective and surjective.*

PROOF:  Suppose $T \in \mathcal{L}(V, W)$. We need to show that $T$ is invertible if and only if it is injective and surjective.

First suppose that $T$ is invertible. To show that $T$ is injective, suppose that $u, v \in V$ and $Tu = Tv$. Then

$$u = T^{-1}(Tu) = T^{-1}(Tv) = v,$$

so $u = v$. Hence $T$ is injective.

We are still assuming that $T$ is invertible. Now we want to prove that $T$ is surjective. To do this, let $w \in W$. Then $w = T(T^{-1}w)$, which shows that $w$ is in the range of $T$. Thus range $T = W$, and hence $T$ is surjective, completing this direction of the proof.

Now suppose that $T$ is injective and surjective. We want to prove that $T$ is invertible. For each $w \in W$, define $Sw$ to be the unique element of $V$ such that $T(Sw) = w$ (the existence and uniqueness of such an element follow from the surjectivity and injectivity of $T$). Clearly $TS$ equals the identity map on $W$. To prove that $ST$ equals the identity map on $V$, let $v \in V$. Then

$$T(STv) = (TS)(Tv) = I(Tv) = Tv.$$

This equation implies that $STv = v$ (because $T$ is injective), and thus $ST$ equals the identity map on $V$. To complete the proof, we need to show that $S$ is linear. To do this, let $w_1, w_2 \in W$. Then

$$T(Sw_1 + Sw_2) = T(Sw_1) + T(Sw_2) = w_1 + w_2.$$

Thus $Sw_1 + Sw_2$ is the unique element of $V$ that $T$ maps to $w_1 + w_2$. By the definition of $S$, this implies that $S(w_1 + w_2) = Sw_1 + Sw_2$. Hence $S$ satisfies the additive property required for linearity. The proof of homogeneity is similar. Specifically, if $w \in W$ and $a \in \mathbf{F}$, then

$$T(aSw) = aT(Sw) = aw.$$

Thus $aSw$ is the unique element of $V$ that $T$ maps to $aw$. By the definition of $S$, this implies that $S(aw) = aSw$. Hence $S$ is linear, as desired. ∎

Two vector spaces are called **isomorphic** if there is an invertible linear map from one vector space onto the other one. As abstract vector spaces, two isomorphic spaces have the same properties. From this viewpoint, you can think of an invertible linear map as a relabeling of the elements of a vector space.

*The Greek word **isos** means equal; the Greek word **morph** means shape. Thus **isomorphic** literally means equal shape.*

If two vector spaces are isomorphic and one of them is finite dimensional, then so is the other one. To see this, suppose that $V$ and $W$ are isomorphic and that $T \in \mathcal{L}(V, W)$ is an invertible linear map. If $V$ is finite dimensional, then so is $W$ (by 3.4). The same reasoning, with $T$ replaced with $T^{-1} \in \mathcal{L}(W, V)$, shows that if $W$ is finite dimensional, then so is $V$. Actually much more is true, as the following theorem shows.

**3.18   Theorem:** *Two finite-dimensional vector spaces are isomorphic if and only if they have the same dimension.*

PROOF:   First suppose $V$ and $W$ are isomorphic finite-dimensional vector spaces. Thus there exists an invertible linear map $T$ from $V$ onto $W$. Because $T$ is invertible, we have $\text{null } T = \{0\}$ and $\text{range } T = W$. Thus $\dim \text{null } T = 0$ and $\dim \text{range } T = \dim W$. The formula

$$\dim V = \dim \text{null } T + \dim \text{range } T$$

(see 3.4) thus becomes the equation $\dim V = \dim W$, completing the proof in one direction.

To prove the other direction, suppose $V$ and $W$ are finite-dimensional vector spaces with the same dimension. Let $(v_1, \ldots, v_n)$ be a basis of $V$ and $(w_1, \ldots, w_n)$ be a basis of $W$. Let $T$ be the linear map from $V$ to $W$ defined by

$$T(a_1 v_1 + \cdots + a_n v_n) = a_1 w_1 + \cdots + a_n w_n.$$

Then $T$ is surjective because $(w_1, \ldots, w_n)$ spans $W$, and $T$ is injective because $(w_1, \ldots, w_n)$ is linearly independent. Because $T$ is injective and surjective, it is invertible (see 3.17), and hence $V$ and $W$ are isomorphic, as desired.                                                                                                 ■

*Because every*
*finite-dimensional*
*vector space is*
*isomorphic to some* **F**$^n$,
*why bother with*
*abstract vector spaces?*
*To answer this*
*question, note that an*
*investigation of* **F**$^n$
*would soon lead to*
*vector spaces that do*
*not equal* **F**$^n$. *For*
*example, we would*
*encounter the null*
*space and range of*
*linear maps, the set of*
*matrices* Mat$(n, n, \mathbf{F})$,
*and the polynomials*
$\mathcal{P}_n(\mathbf{F})$. *Though each of*
*these vector spaces is*
*isomorphic to some*
**F**$^m$, *thinking of them*
*that way often adds*
*complexity but no new*
*insight.*

The last theorem implies that every finite-dimensional vector space is isomorphic to some $\mathbf{F}^n$. Specifically, if $V$ is a finite-dimensional vector space and dim $V = n$, then $V$ and $\mathbf{F}^n$ are isomorphic.

If $(v_1, \ldots, v_n)$ is a basis of $V$ and $(w_1, \ldots, w_m)$ is a basis of $W$, then for each $T \in \mathcal{L}(V, W)$, we have a matrix $\mathcal{M}(T) \in \mathrm{Mat}(m, n, \mathbf{F})$. In other words, once bases have been fixed for $V$ and $W$, $\mathcal{M}$ becomes a function from $\mathcal{L}(V, W)$ to $\mathrm{Mat}(m, n, \mathbf{F})$. Notice that 3.9 and 3.10 show that $\mathcal{M}$ is a linear map. This linear map is actually invertible, as we now show.

**3.19    Proposition:**    *Suppose that* $(v_1, \ldots, v_n)$ *is a basis of* $V$ *and* $(w_1, \ldots, w_m)$ *is a basis of* $W$. *Then* $\mathcal{M}$ *is an invertible linear map between* $\mathcal{L}(V, W)$ *and* $\mathrm{Mat}(m, n, \mathbf{F})$.

PROOF:  We have already noted that $\mathcal{M}$ is linear, so we need only prove that $\mathcal{M}$ is injective and surjective (by 3.17). Both are easy. Let's begin with injectivity. If $T \in \mathcal{L}(V, W)$ and $\mathcal{M}(T) = 0$, then $Tv_k = 0$ for $k = 1, \ldots, n$. Because $(v_1, \ldots, v_n)$ is a basis of $V$, this implies that $T = 0$. Thus $\mathcal{M}$ is injective (by 3.2).

To prove that $\mathcal{M}$ is surjective, let

$$A = \begin{bmatrix} a_{1,1} & \cdots & a_{1,n} \\ \vdots & & \vdots \\ a_{m,1} & \cdots & a_{m,n} \end{bmatrix}$$

be a matrix in $\mathrm{Mat}(m, n, \mathbf{F})$. Let $T$ be the linear map from $V$ to $W$ such that

$$Tv_k = \sum_{j=1}^{m} a_{j,k} w_j$$

for $k = 1, \ldots, n$. Obviously $\mathcal{M}(T)$ equals $A$, and so the range of $\mathcal{M}$ equals $\mathrm{Mat}(m, n, \mathbf{F})$, as desired.                                      ■

An obvious basis of $\mathrm{Mat}(m, n, \mathbf{F})$ consists of those $m$-by-$n$ matrices that have 0 in all entries except for a 1 in one entry. There are $mn$ such matrices, so the dimension of $\mathrm{Mat}(m, n, \mathbf{F})$ equals $mn$.

Now we can determine the dimension of the vector space of linear maps from one finite-dimensional vector space to another.

**3.20  Proposition:** *If $V$ and $W$ are finite dimensional, then $\mathcal{L}(V, W)$ is finite dimensional and*

$$\dim \mathcal{L}(V, W) = (\dim V)(\dim W).$$

PROOF:  This follows from the equation $\dim \operatorname{Mat}(m, n, \mathbf{F}) = mn$, 3.18, and 3.19.    ∎

A linear map from a vector space to itself is called an ***operator***. If we want to specify the vector space, we say that a linear map $T: V \to V$ is an operator on $V$. Because we are so often interested in linear maps from a vector space into itself, we use the notation $\mathcal{L}(V)$ to denote the set of all operators on $V$. In other words, $\mathcal{L}(V) = \mathcal{L}(V, V)$.

Recall from 3.17 that a linear map is invertible if it is injective and surjective. For a linear map of a vector space into itself, you might wonder whether injectivity alone, or surjectivity alone, is enough to imply invertibility. On infinite-dimensional vector spaces neither condition alone implies invertibility. We can see this from some examples we have already considered. The multiplication by $x^2$ operator (from $\mathcal{P}(\mathbf{R})$ to itself) is injective but not surjective. The backward shift (from $\mathbf{F}^\infty$ to itself) is surjective but not injective. In view of these examples, the next theorem is remarkable—it states that for maps from a finite-dimensional vector space to itself, either injectivity or surjectivity alone implies the other condition.

*The deepest and most important parts of linear algebra, as well as most of the rest of this book, deal with operators.*

**3.21  Theorem:** *Suppose $V$ is finite dimensional. If $T \in \mathcal{L}(V)$, then the following are equivalent:*

(a)    *$T$ is invertible;*

(b)    *$T$ is injective;*

(c)    *$T$ is surjective.*

PROOF:  Suppose $T \in \mathcal{L}(V)$. Clearly (a) implies (b).

Now suppose (b) holds, so that $T$ is injective. Thus $\operatorname{null} T = \{0\}$ (by 3.2). From 3.4 we have

$$\begin{aligned}\dim \operatorname{range} T &= \dim V - \dim \operatorname{null} T \\ &= \dim V,\end{aligned}$$

which implies that range $T$ equals $V$ (see Exercise 11 in Chapter 2). Thus $T$ is surjective. Hence (b) implies (c).

Now suppose (c) holds, so that $T$ is surjective. Thus range $T = V$. From 3.4 we have

$$\dim \operatorname{null} T = \dim V - \dim \operatorname{range} T$$
$$= 0,$$

which implies that null $T$ equals $\{0\}$. Thus $T$ is injective (by 3.2), and so $T$ is invertible (we already knew that $T$ was surjective). Hence (c) implies (a), completing the proof.                                    ■

# Exercises

1.  Show that every linear map from a one-dimensional vector space to itself is multiplication by some scalar. More precisely, prove that if $\dim V = 1$ and $T \in \mathcal{L}(V, V)$, then there exists $a \in \mathbf{F}$ such that $Tv = av$ for all $v \in V$.

2.  Give an example of a function $f : \mathbf{R}^2 \to \mathbf{R}$ such that

    $$f(av) = af(v)$$

    for all $a \in \mathbf{R}$ and all $v \in \mathbf{R}^2$ but $f$ is not linear.

    *Exercise 2 shows that homogeneity alone is not enough to imply that a function is a linear map. Additivity alone is also not enough to imply that a function is a linear map, although the proof of this involves advanced tools that are beyond the scope of this book.*

3.  Suppose that $V$ is finite dimensional. Prove that any linear map on a subspace of $V$ can be extended to a linear map on $V$. In other words, show that if $U$ is a subspace of $V$ and $S \in \mathcal{L}(U, W)$, then there exists $T \in \mathcal{L}(V, W)$ such that $Tu = Su$ for all $u \in U$.

4.  Suppose that $T$ is a linear map from $V$ to $\mathbf{F}$. Prove that if $u \in V$ is not in $\operatorname{null} T$, then

    $$V = \operatorname{null} T \oplus \{au : a \in \mathbf{F}\}.$$

5.  Suppose that $T \in \mathcal{L}(V, W)$ is injective and $(v_1, \ldots, v_n)$ is linearly independent in $V$. Prove that $(Tv_1, \ldots, Tv_n)$ is linearly independent in $W$.

6.  Prove that if $S_1, \ldots, S_n$ are injective linear maps such that $S_1 \ldots S_n$ makes sense, then $S_1 \ldots S_n$ is injective.

7.  Prove that if $(v_1, \ldots, v_n)$ spans $V$ and $T \in \mathcal{L}(V, W)$ is surjective, then $(Tv_1, \ldots, Tv_n)$ spans $W$.

8.  Suppose that $V$ is finite dimensional and that $T \in \mathcal{L}(V, W)$. Prove that there exists a subspace $U$ of $V$ such that $U \cap \operatorname{null} T = \{0\}$ and $\operatorname{range} T = \{Tu : u \in U\}$.

9.  Prove that if $T$ is a linear map from $\mathbf{F}^4$ to $\mathbf{F}^2$ such that

    $$\operatorname{null} T = \{(x_1, x_2, x_3, x_4) \in \mathbf{F}^4 : x_1 = 5x_2 \text{ and } x_3 = 7x_4\},$$

    then $T$ is surjective.

10.  Prove that there does not exist a linear map from $\mathbf{F}^5$ to $\mathbf{F}^2$ whose null space equals

$$\{(x_1, x_2, x_3, x_4, x_5) \in \mathbf{F}^5 : x_1 = 3x_2 \text{ and } x_3 = x_4 = x_5\}.$$

11.  Prove that if there exists a linear map on $V$ whose null space and range are both finite dimensional, then $V$ is finite dimensional.

12.  Suppose that $V$ and $W$ are both finite dimensional. Prove that there exists a surjective linear map from $V$ onto $W$ if and only if $\dim W \le \dim V$.

13.  Suppose that $V$ and $W$ are finite dimensional and that $U$ is a subspace of $V$. Prove that there exists $T \in \mathcal{L}(V, W)$ such that $\operatorname{null} T = U$ if and only if $\dim U \ge \dim V - \dim W$.

14.  Suppose that $W$ is finite dimensional and $T \in \mathcal{L}(V, W)$. Prove that $T$ is injective if and only if there exists $S \in \mathcal{L}(W, V)$ such that $ST$ is the identity map on $V$.

15.  Suppose that $V$ is finite dimensional and $T \in \mathcal{L}(V, W)$. Prove that $T$ is surjective if and only if there exists $S \in \mathcal{L}(W, V)$ such that $TS$ is the identity map on $W$.

16.  Suppose that $U$ and $V$ are finite-dimensional vector spaces and that $S \in \mathcal{L}(V, W)$, $T \in \mathcal{L}(U, V)$. Prove that

$$\dim \operatorname{null} ST \le \dim \operatorname{null} S + \dim \operatorname{null} T.$$

17.  Prove that the distributive property holds for matrix addition and matrix multiplication. In other words, suppose $A$, $B$, and $C$ are matrices whose sizes are such that $A(B + C)$ makes sense. Prove that $AB + AC$ makes sense and that $A(B + C) = AB + AC$.

18.  Prove that matrix multiplication is associative. In other words, suppose $A$, $B$, and $C$ are matrices whose sizes are such that $(AB)C$ makes sense. Prove that $A(BC)$ makes sense and that $(AB)C = A(BC)$.

19.    Suppose $T \in \mathcal{L}(\mathbf{F}^n, \mathbf{F}^m)$ and that

*This exercise shows
that T has the form
promised on page 39.*

$$\mathcal{M}(T) = \begin{bmatrix} a_{1,1} & \cdots & a_{1,n} \\ \vdots & & \vdots \\ a_{m,1} & \cdots & a_{m,n} \end{bmatrix},$$

where we are using the standard bases. Prove that

$$T(x_1, \ldots, x_n) = (a_{1,1}x_1 + \cdots + a_{1,n}x_n, \ldots, a_{m,1}x_1 + \cdots + a_{m,n}x_n)$$

for every $(x_1, \ldots, x_n) \in \mathbf{F}^n$.

20.    Suppose $(v_1, \ldots, v_n)$ is a basis of $V$. Prove that the function $T : V \to \mathrm{Mat}(n, 1, \mathbf{F})$ defined by

$$Tv = \mathcal{M}(v)$$

is an invertible linear map of $V$ onto $\mathrm{Mat}(n, 1, \mathbf{F})$; here $\mathcal{M}(v)$ is the matrix of $v \in V$ with respect to the basis $(v_1, \ldots, v_n)$.

21.    Prove that every linear map from $\mathrm{Mat}(n, 1, \mathbf{F})$ to $\mathrm{Mat}(m, 1, \mathbf{F})$ is given by a matrix multiplication. In other words, prove that if $T \in \mathcal{L}(\mathrm{Mat}(n, 1, \mathbf{F}), \mathrm{Mat}(m, 1, \mathbf{F}))$, then there exists an $m$-by-$n$ matrix $A$ such that $TB = AB$ for every $B \in \mathrm{Mat}(n, 1, \mathbf{F})$.

22.    Suppose that $V$ is finite dimensional and $S, T \in \mathcal{L}(V)$. Prove that $ST$ is invertible if and only if both $S$ and $T$ are invertible.

23.    Suppose that $V$ is finite dimensional and $S, T \in \mathcal{L}(V)$. Prove that $ST = I$ if and only if $TS = I$.

24.    Suppose that $V$ is finite dimensional and $T \in \mathcal{L}(V)$. Prove that $T$ is a scalar multiple of the identity if and only if $ST = TS$ for every $S \in \mathcal{L}(V)$.

25.    Prove that if $V$ is finite dimensional with $\dim V > 1$, then the set of noninvertible operators on $V$ is not a subspace of $\mathcal{L}(V)$.

26.    Suppose $n$ is a positive integer and $a_{i,j} \in \mathbf{F}$ for $i, j = 1, \ldots, n$. Prove that the following are equivalent:

(a)    The trivial solution $x_1 = \cdots = x_n = 0$ is the only solution to the homogeneous system of equations

$$\sum_{k=1}^{n} a_{1,k} x_k = 0$$

$$\vdots$$

$$\sum_{k=1}^{n} a_{n,k} x_k = 0.$$

(b)    For every $c_1, \ldots, c_n \in \mathbf{F}$, there exists a solution to the system of equations

$$\sum_{k=1}^{n} a_{1,k} x_k = c_1$$

$$\vdots$$

$$\sum_{k=1}^{n} a_{n,k} x_k = c_n.$$

Note that here we have the same number of equations as variables.

# CHAPTER 4

# *Polynomials*

This short chapter contains no linear algebra. It does contain the background material on polynomials that we will need in our study of linear maps from a vector space to itself. Many of the results in this chapter will already be familiar to you from other courses; they are included here for completeness. Because this chapter is not about linear algebra, your instructor may go through it rapidly. You may not be asked to scrutinize all the proofs. Make sure, however, that you at least read and understand the statements of all the results in this chapter—they will be used in the rest of the book.

Recall that **F** denotes **R** or **C**.

# Degree

Recall that a function $p : \mathbf{F} \to \mathbf{F}$ is called a polynomial with coefficients in $\mathbf{F}$ if there exist $a_0, \ldots, a_m \in \mathbf{F}$ such that

$$p(z) = a_0 + a_1 z + a_2 z^2 + \cdots + a_m z^m$$

for all $z \in \mathbf{F}$. If $p$ can be written in the form above with $a_m \neq 0$, then we say that $p$ has degree $m$. If all the coefficients $a_0, \ldots, a_m$ equal 0, then we say that $p$ has degree $-\infty$. For all we know at this stage, a polynomial may have more than one degree because we have not yet proved that the coefficients in the equation above are uniquely determined by the function $p$.

*When necessary, use the obvious arithmetic with $-\infty$. For example, $-\infty < m$ and $-\infty + m = -\infty$ for every integer $m$. The 0 polynomial is declared to have degree $-\infty$ so that exceptions are not needed for various reasonable results. For example, the degree of $pq$ equals the degree of $p$ plus the degree of $q$ even if $p = 0$.*

Recall that $\mathcal{P}(\mathbf{F})$ denotes the vector space of all polynomials with coefficients in $\mathbf{F}$ and that $\mathcal{P}_m(\mathbf{F})$ is the subspace of $\mathcal{P}(\mathbf{F})$ consisting of the polynomials with coefficients in $\mathbf{F}$ and degree at most $m$. A number $\lambda \in \mathbf{F}$ is called a **root** of a polynomial $p \in \mathcal{P}(\mathbf{F})$ if

$$p(\lambda) = 0.$$

Roots play a crucial role in the study of polynomials. We begin by showing that $\lambda$ is a root of $p$ if and only if $p$ is a polynomial multiple of $z - \lambda$.

**4.1**   **Proposition:**   *Suppose $p \in \mathcal{P}(\mathbf{F})$ is a polynomial with degree $m \geq 1$. Let $\lambda \in \mathbf{F}$. Then $\lambda$ is a root of $p$ if and only if there is a polynomial $q \in \mathcal{P}(\mathbf{F})$ with degree $m - 1$ such that*

**4.2**                          $$p(z) = (z - \lambda)q(z)$$

*for all $z \in \mathbf{F}$.*

PROOF: One direction is obvious. Namely, suppose there is a polynomial $q \in \mathcal{P}(\mathbf{F})$ such that 4.2 holds. Then

$$p(\lambda) = (\lambda - \lambda)q(\lambda) = 0,$$

and hence $\lambda$ is a root of $p$, as desired.

To prove the other direction, suppose that $\lambda \in \mathbf{F}$ is a root of $p$. Let $a_0, \ldots, a_m \in \mathbf{F}$ be such that $a_m \neq 0$ and

$$p(z) = a_0 + a_1 z + a_2 z^2 + \cdots + a_m z^m$$

for all $z \in \mathbf{F}$. Because $p(\lambda) = 0$, we have

$$0 = a_0 + a_1\lambda + a_2\lambda^2 + \cdots + a_m\lambda^m.$$

Subtracting the last two equations, we get

$$p(z) = a_1(z - \lambda) + a_2(z^2 - \lambda^2) + \cdots + a_m(z^m - \lambda^m)$$

for all $z \in \mathbf{F}$. For each $j = 2, \ldots, m$, we can write

$$z^j - \lambda^j = (z - \lambda)q_{j-1}(z)$$

for all $z \in \mathbf{F}$, where $q_{j-1}$ is a polynomial with degree $j - 1$ (specifically, take $q_{j-1}(z) = z^{j-1} + z^{j-2}\lambda + \cdots + z\lambda^{j-2} + \lambda^{j-1}$). Thus

$$p(z) = (z - \lambda) \underbrace{(a_1 + a_2q_2(z) + \cdots + a_mq_{m-1}(z))}_{q(z)}$$

for all $z \in \mathbf{F}$. Clearly $q$ is a polynomial with degree $m - 1$, as desired. ∎

Now we can prove that polynomials do not have too many roots.

**4.3**    **Corollary:** *Suppose* $p \in \mathcal{P}(\mathbf{F})$ *is a polynomial with degree* $m \geq 0$. *Then* $p$ *has at most* $m$ *distinct roots in* $\mathbf{F}$.

PROOF:   If $m = 0$, then $p(z) = a_0 \neq 0$ and so $p$ has no roots. If $m = 1$, then $p(z) = a_0 + a_1z$, with $a_1 \neq 0$, and $p$ has exactly one root, namely, $-a_0/a_1$. Now suppose $m > 1$. We use induction on $m$, assuming that every polynomial with degree $m - 1$ has at most $m - 1$ distinct roots. If $p$ has no roots in $\mathbf{F}$, then we are done. If $p$ has a root $\lambda \in \mathbf{F}$, then by 4.1 there is a polynomial $q$ with degree $m - 1$ such that

$$p(z) = (z - \lambda)q(z)$$

for all $z \in \mathbf{F}$. The equation above shows that if $p(z) = 0$, then either $z = \lambda$ or $q(z) = 0$. In other words, the roots of $p$ consist of $\lambda$ and the roots of $q$. By our induction hypothesis, $q$ has at most $m - 1$ distinct roots in $\mathbf{F}$. Thus $p$ has at most $m$ distinct roots in $\mathbf{F}$.          ∎

The next result states that if a polynomial is identically 0, then all its coefficients must be 0.

**4.4**    **Corollary:**  *Suppose $a_0, \ldots, a_m \in \mathbf{F}$. If*

$$a_0 + a_1 z + a_2 z^2 + \cdots + a_m z^m = 0$$

*for all $z \in \mathbf{F}$, then $a_0 = \cdots = a_m = 0$.*

PROOF:  Suppose $a_0 + a_1 z + a_2 z^2 + \cdots + a_m z^m$ equals 0 for all $z \in \mathbf{F}$. By 4.3, no nonnegative integer can be the degree of this polynomial. Thus all the coefficients equal 0.  ∎

The corollary above implies that $(1, z, \ldots, z^m)$ is linearly independent in $\mathcal{P}(\mathbf{F})$ for every nonnegative integer $m$. We had noted this earlier (in Chapter 2), but now we have a complete proof. This linear independence implies that each polynomial can be represented in only one way as a linear combination of functions of the form $z^j$. In particular, the degree of a polynomial is unique.

If $p$ and $q$ are nonnegative integers, with $p \neq 0$, then there exist nonnegative integers $s$ and $r$ such that

$$q = sp + r.$$

and $r < p$. Think of dividing $q$ by $p$, getting $s$ with remainder $r$. Our next task is to prove an analogous result for polynomials.

Let $\deg p$ denote the degree of a polynomial $p$. The next result is often called the division algorithm, though as stated here it is not really an algorithm, just a useful lemma.

*Think of 4.6 as giving the remainder $r$ when $q$ is divided by $p$.*    **4.5**    **Division Algorithm:**  *Suppose $p, q \in \mathcal{P}(\mathbf{F})$, with $p \neq 0$. Then there exist polynomials $s, r \in \mathcal{P}(\mathbf{F})$ such that*

**4.6**                                     $$q = sp + r$$

*and $\deg r < \deg p$.*

PROOF:   Choose $s \in \mathcal{P}(\mathbf{F})$ such that $q - sp$ has degree as small as possible. Let $r = q - sp$. Thus 4.6 holds, and all that remains is to show that $\deg r < \deg p$. Suppose that $\deg r \geq \deg p$. If $c \in \mathbf{F}$ and $j$ is a nonnegative integer, then

$$q - (s + cz^j)p = r - cz^j p.$$

Choose $j$ and $c$ so that the polynomial on the right side of this equation has degree less than $\deg r$ (specifically, take $j = \deg r - \deg p$ and then choose $c$ so that the coefficients of $z^{\deg r}$ in $r$ and in $cz^j p$ are equal). This contradicts our choice of $s$ as the polynomial that produces the smallest degree for expressions of the form $q - sp$, completing the proof.                                                                ∎

## Complex Coefficients

So far we have been handling polynomials with complex coefficients and polynomials with real coefficients simultaneously through our convention that **F** denotes **R** or **C**. Now we will see some differences between these two cases. In this section we treat polynomials with complex coefficients. In the next section we will use our results about polynomials with complex coefficients to prove corresponding results for polynomials with real coefficients.

Though this chapter contains no linear algebra, the results so far have nonetheless been proved using algebra. The next result, though called the fundamental theorem of algebra, requires analysis for its proof. The short proof presented here uses tools from complex analysis. If you have not had a course in complex analysis, this proof will almost certainly be meaningless to you. In that case, just accept the fundamental theorem of algebra as something that we need to use but whose proof requires more advanced tools that you may learn in later courses.

**4.7    Fundamental Theorem of Algebra:** *Every nonconstant polynomial with complex coefficients has a root.*

PROOF:    Let $p$ be a nonconstant polynomial with complex coefficients. Suppose that $p$ has no roots. Then $1/p$ is an analytic function on **C**. Furthermore, $p(z) \to \infty$ as $z \to \infty$, which implies that $1/p \to 0$ as $z \to \infty$. Thus $1/p$ is a bounded analytic function on **C**. By Liouville's theorem, any such function must be constant. But if $1/p$ is constant, then $p$ is constant, contradicting our assumption that $p$ is nonconstant.    ∎

The fundamental theorem of algebra leads to the following factorization result for polynomials with complex coefficients. Note that in this factorization, the numbers $\lambda_1, \ldots, \lambda_m$ are precisely the roots of $p$, for these are the only values of $z$ for which the right side of 4.9 equals 0.

*This is an existence theorem. The quadratic formula gives the roots explicitly for polynomials of degree 2. Similar but more complicated formulas exist for polynomials of degree 3 and 4. No such formulas exist for polynomials of degree 5 and above.*

**4.8     Corollary:**  *If $p \in \mathcal{P}(\mathbf{C})$ is a nonconstant polynomial, then $p$
has a unique factorization (except for the order of the factors) of the
form*

**4.9**                       $$p(z) = c(z - \lambda_1) \dots (z - \lambda_m),$$

*where $c, \lambda_1, \dots, \lambda_m \in \mathbf{C}$.*

PROOF:   Let $p \in \mathcal{P}(\mathbf{C})$ and let $m$ denote the degree of $p$. We will use
induction on $m$. If $m = 1$, then clearly the desired factorization exists
and is unique. So assume that $m > 1$ and that the desired factorization
exists and is unique for all polynomials of degree $m - 1$.

First we will show that the desired factorization of $p$ exists. By the
fundamental theorem of algebra (4.7), $p$ has a root $\lambda$. By 4.1, there is a
polynomial $q$ with degree $m - 1$ such that

$$p(z) = (z - \lambda)q(z)$$

for all $z \in \mathbf{C}$. Our induction hypothesis implies that $q$ has the desired
factorization, which when plugged into the equation above gives the
desired factorization of $p$.

Now we turn to the question of uniqueness. Clearly $c$ is uniquely
determined by 4.9—it must equal the coefficient of $z^m$ in $p$. So we need
only show that except for the order, there is only one way to choose
$\lambda_1, \dots, \lambda_m$. If

$$(z - \lambda_1) \dots (z - \lambda_m) = (z - \tau_1) \dots (z - \tau_m)$$

for all $z \in \mathbf{C}$, then because the left side of the equation above equals 0
when $z = \lambda_1$, one of the $\tau$'s on the right side must equal $\lambda_1$. Relabeling,
we can assume that $\tau_1 = \lambda_1$. Now for $z \neq \lambda_1$, we can divide both sides
of the equation above by $z - \lambda_1$, getting

$$(z - \lambda_2) \dots (z - \lambda_m) = (z - \tau_2) \dots (z - \tau_m)$$

for all $z \in \mathbf{C}$ except possibly $z = \lambda_1$. Actually the equation above
must hold for all $z \in \mathbf{C}$ because otherwise by subtracting the right side
from the left side we would get a nonzero polynomial that has infinitely
many roots. The equation above and our induction hypothesis imply
that except for the order, the $\lambda$'s are the same as the $\tau$'s, completing
the proof of the uniqueness.                                          ∎

# Real Coefficients

Before discussing polynomials with real coefficients, we need to learn a bit more about the complex numbers.

Suppose $z = a + bi$, where $a$ and $b$ are real numbers. Then $a$ is called the **real part** of $z$, denoted $\operatorname{Re} z$, and $b$ is called the **imaginary part** of $z$, denoted $\operatorname{Im} z$. Thus for every complex number $z$, we have

$$z = \operatorname{Re} z + (\operatorname{Im} z)i.$$

The **complex conjugate** of $z \in \mathbf{C}$, denoted $\bar{z}$, is defined by

$$\bar{z} = \operatorname{Re} z - (\operatorname{Im} z)i.$$

*Note that $z = \bar{z}$ if and only if $z$ is a real number.*

For example, $\overline{2 + 3i} = 2 - 3i$.

The **absolute value** of a complex number $z$, denoted $|z|$, is defined by

$$|z| = \sqrt{(\operatorname{Re} z)^2 + (\operatorname{Im} z)^2}.$$

For example, $|1 + 2i| = \sqrt{5}$. Note that $|z|$ is always a nonnegative number.

You should verify that the real and imaginary parts, absolute value, and complex conjugate have the following properties:

**additivity of real part**
$\operatorname{Re}(w + z) = \operatorname{Re} w + \operatorname{Re} z$ for all $w, z \in \mathbf{C}$;

**additivity of imaginary part**
$\operatorname{Im}(w + z) = \operatorname{Im} w + \operatorname{Im} z$ for all $w, z \in \mathbf{C}$;

**sum of $z$ and $\bar{z}$**
$z + \bar{z} = 2 \operatorname{Re} z$ for all $z \in \mathbf{C}$;

**difference of $z$ and $\bar{z}$**
$z - \bar{z} = 2(\operatorname{Im} z)i$ for all $z \in \mathbf{C}$;

**product of $z$ and $\bar{z}$**
$z\bar{z} = |z|^2$ for all $z \in \mathbf{C}$;

**additivity of complex conjugate**
$\overline{w + z} = \bar{w} + \bar{z}$ for all $w, z \in \mathbf{C}$;

**multiplicativity of complex conjugate**
$\overline{wz} = \bar{w}\bar{z}$ for all $w, z \in \mathbf{C}$;

**conjugate of conjugate**
$\bar{\bar{z}} = z$ for all $z \in \mathbf{C}$;

**multiplicativity of absolute value**
$|wz| = |w|\,|z|$ for all $w, z \in C$.

In the next result, we need to think of a polynomial with real coefficients as an element of $\mathcal{P}(\mathbf{C})$. This makes sense because every real number is also a complex number.

*A polynomial with real coefficients may have no real roots. For example, the polynomial $1 + x^2$ has no real roots. The failure of the fundamental theorem of algebra for $\mathbf{R}$ accounts for the differences between operators on real and complex vector spaces, as we will see in later chapters.*

**4.10   Proposition:**  *Suppose $p$ is a polynomial with real coefficients. If $\lambda \in \mathbf{C}$ is a root of $p$, then so is $\bar{\lambda}$.*

PROOF:   Let
$$p(z) = a_0 + a_1 z + \cdots + a_m z^m,$$
where $a_0, \ldots, a_m$ are real numbers. Suppose $\lambda \in \mathbf{C}$ is a root of $p$. Then
$$a_0 + a_1\lambda + \cdots + a_m\lambda^m = 0.$$
Take the complex conjugate of both sides of this equation, obtaining
$$a_0 + a_1\bar{\lambda} + \cdots + a_m\bar{\lambda}^m = 0,$$
where we have used some of the basic properties of complex conjugation listed earlier. The equation above shows that $\bar{\lambda}$ is a root of $p$.   ∎

We want to prove a factorization theorem for polynomials with real coefficients. To do this, we begin by characterizing the polynomials with real coefficients and degree 2 that can be written as the product of two polynomials with real coefficients and degree 1.

*Think about the connection between the quadratic formula and this proposition.*

**4.11   Proposition:**  *Let $\alpha, \beta \in \mathbf{R}$. Then there is a polynomial factorization of the form*

**4.12**  $$x^2 + \alpha x + \beta = (x - \lambda_1)(x - \lambda_2),$$

*with $\lambda_1, \lambda_2 \in \mathbf{R}$, if and only if $\alpha^2 \geq 4\beta$.*

PROOF:   Notice that

**4.13**  $$x^2 + \alpha x + \beta = (x + \frac{\alpha}{2})^2 + (\beta - \frac{\alpha^2}{4}).$$

First suppose that $\alpha^2 < 4\beta$. Then clearly the right side of the equation above is positive for every $x \in \mathbf{R}$, and hence the polynomial $x^2 + \alpha x + \beta$ has no real roots. Thus no factorization of the form 4.12, with $\lambda_1, \lambda_2 \in \mathbf{R}$, can exist.

Conversely, now suppose that $\alpha^2 \geq 4\beta$. Thus there is a real number $c$ such that $c^2 = \frac{\alpha^2}{4} - \beta$. From 4.13, we have

$$x^2 + \alpha x + \beta = (x + \frac{\alpha}{2})^2 - c^2$$
$$= (x + \frac{\alpha}{2} + c)(x + \frac{\alpha}{2} - c),$$

which gives the desired factorization.                                       ■

In the following theorem, each term of the form $x^2 + \alpha_j x + \beta_j$, with $\alpha_j{}^2 < 4\beta_j$, cannot be factored into the product of two polynomials with real coefficients and degree 1 (by 4.11). Note that in the factorization below, the numbers $\lambda_1, \ldots, \lambda_m$ are precisely the real roots of $p$, for these are the only real values of $x$ for which the right side of the equation below equals 0.

**4.14   Theorem:**  *If $p \in \mathcal{P}(\mathbf{R})$ is a nonconstant polynomial, then $p$ has a unique factorization (except for the order of the factors) of the form*

$$p(x) = c(x - \lambda_1) \ldots (x - \lambda_m)(x^2 + \alpha_1 x + \beta_1) \ldots (x^2 + \alpha_M x + \beta_M),$$

*where $c, \lambda_1, \ldots, \lambda_m \in \mathbf{R}$ and $(\alpha_1, \beta_1), \ldots, (\alpha_M, \beta_M) \in \mathbf{R}^2$ with $\alpha_j{}^2 < 4\beta_j$ for each $j$.*

*Here either $m$ or $M$ may equal 0.*

PROOF:   Let $p \in \mathcal{P}(\mathbf{R})$ be a nonconstant polynomial. We can think of $p$ as an element of $\mathcal{P}(\mathbf{C})$ (because every real number is a complex number). The idea of the proof is to use the factorization 4.8 of $p$ as a polynomial with complex coefficients. Complex but nonreal roots of $p$ come in pairs; see 4.10. Thus if the factorization of $p$ as an element of $\mathcal{P}(\mathbf{C})$ includes terms of the form $(x - \lambda)$ with $\lambda$ a nonreal complex number, then $(x - \bar{\lambda})$ is also a term in the factorization. Combining these two terms, we get a quadratic term of the required form.

The idea sketched in the paragraph above almost provides a proof of the existence of our desired factorization. However, we need to be careful about one point. Suppose $\lambda$ is a nonreal complex number and $(x - \lambda)$ is a term in the factorization of $p$ as an element of $\mathcal{P}(\mathbf{C})$. We are guaranteed by 4.10 that $(x - \bar{\lambda})$ also appears as a term in the factorization, but 4.10 does not state that these two factors appear the same number of times, as needed to make the idea above work. However, all is well. We can write

$$p(x) = (x - \lambda)(x - \bar{\lambda})q(x)$$
$$= (x^2 - 2(\operatorname{Re}\lambda)x + |\lambda|^2)q(x)$$

for some polynomial $q \in \mathcal{P}(\mathbf{C})$ with degree two less than the degree of $p$. If we can prove that $q$ has real coefficients, then, by using induction on the degree of $p$, we can conclude that $(x - \lambda)$ appears in the factorization of $p$ exactly as many times as $(x - \bar{\lambda})$.

To prove that $q$ has real coefficients, we solve the equation above for $q$, getting

$$q(x) = \frac{p(x)}{x^2 - 2(\operatorname{Re}\lambda)x + |\lambda|^2}$$

*Here we are not dividing by 0 because the roots of $x^2 - 2(\operatorname{Re}\lambda)x + |\lambda|^2$ are $\lambda$ and $\bar{\lambda}$, neither of which is real.*

for all $x \in \mathbf{R}$. The equation above implies that $q(x) \in \mathbf{R}$ for all $x \in \mathbf{R}$. Writing

$$q(x) = a_0 + a_1 x + \cdots + a_{n-2} x^{n-2},$$

where $a_0, \ldots, a_{n-2} \in \mathbf{C}$, we thus have

$$0 = \operatorname{Im} q(x) = (\operatorname{Im} a_0) + (\operatorname{Im} a_1)x + \cdots + (\operatorname{Im} a_{n-2})x^{n-2}$$

for all $x \in \mathbf{R}$. This implies that $\operatorname{Im} a_0, \ldots, \operatorname{Im} a_{n-2}$ all equal 0 (by 4.4). Thus all the coefficients of $q$ are real, as desired, and hence the desired factorization exists.

Now we turn to the question of uniqueness of our factorization. A factor of $p$ of the form $x^2 + \alpha x + \beta$ with $\alpha^2 < 4\beta$ can be uniquely written as $(x - \lambda)(x - \bar{\lambda})$ with $\lambda \in \mathbf{C}$. A moment's thought shows that two different factorizations of $p$ as an element of $\mathcal{P}(\mathbf{R})$ would lead to two different factorizations of $p$ as an element of $\mathcal{P}(\mathbf{C})$, contradicting 4.8. ∎

# Exercises

1.  Suppose $m$ and $n$ are positive integers with $m \le n$. Prove that there exists a polynomial $p \in \mathcal{P}_n(\mathbf{F})$ with exactly $m$ distinct roots.

2.  Suppose that $z_1, \ldots, z_{m+1}$ are distinct elements of $\mathbf{F}$ and that $w_1, \ldots, w_{m+1} \in \mathbf{F}$. Prove that there exists a unique polynomial $p \in \mathcal{P}_m(\mathbf{F})$ such that
    $$p(z_j) = w_j$$
    for $j = 1, \ldots, m + 1$.

3.  Prove that if $p, q \in \mathcal{P}(\mathbf{F})$, with $p \ne 0$, then there exist unique polynomials $s, r \in \mathcal{P}(\mathbf{F})$ such that

    $$q = sp + r$$

    and $\deg r < \deg p$. In other words, add a uniqueness statement to the division algorithm (4.5).

4.  Suppose $p \in \mathcal{P}(\mathbf{C})$ has degree $m$. Prove that $p$ has $m$ distinct roots if and only if $p$ and its derivative $p'$ have no roots in common.

5.  Prove that every polynomial with odd degree and real coefficients has a real root.

# CHAPTER 5

# *Eigenvalues and Eigenvectors*

In Chapter 3 we studied linear maps from one vector space to another vector space. Now we begin our investigation of linear maps from a vector space to itself. Their study constitutes the deepest and most important part of linear algebra. Most of the key results in this area do not hold for infinite-dimensional vector spaces, so we work only on finite-dimensional vector spaces. To avoid trivialities we also want to eliminate the vector space {0} from consideration. Thus we make the following assumption:

> Recall that **F** denotes **R** or **C**.
> Let's agree that for the rest of the book
> *V* will denote a finite-dimensional, nonzero vector space over **F**.

# Invariant Subspaces

In this chapter we develop the tools that will help us understand the structure of operators. Recall that an operator is a linear map from a vector space to itself. Recall also that we denote the set of operators on $V$ by $\mathcal{L}(V)$; in other words, $\mathcal{L}(V) = \mathcal{L}(V, V)$.

Let's see how we might better understand what an operator looks like. Suppose $T \in \mathcal{L}(V)$. If we have a direct sum decomposition

**5.1**
$$V = U_1 \oplus \cdots \oplus U_m,$$

where each $U_j$ is a proper subspace of $V$, then to understand the behavior of $T$, we need only understand the behavior of each $T|_{U_j}$; here $T|_{U_j}$ denotes the restriction of $T$ to the smaller domain $U_j$. Dealing with $T|_{U_j}$ should be easier than dealing with $T$ because $U_j$ is a smaller vector space than $V$. However, if we intend to apply tools useful in the study of operators (such as taking powers), then we have a problem: $T|_{U_j}$ may not map $U_j$ into itself; in other words, $T|_{U_j}$ may not be an operator on $U_j$. Thus we are led to consider only decompositions of the form 5.1 where $T$ maps each $U_j$ into itself.

The notion of a subspace that gets mapped into itself is sufficiently important to deserve a name. Thus, for $T \in \mathcal{L}(V)$ and $U$ a subspace of $V$, we say that $U$ is ***invariant*** under $T$ if $u \in U$ implies $Tu \in U$. In other words, $U$ is invariant under $T$ if $T|_U$ is an operator on $U$. For example, if $T$ is the operator of differentiation on $\mathcal{P}_7(\mathbf{R})$, then $\mathcal{P}_4(\mathbf{R})$ (which is a subspace of $\mathcal{P}_7(\mathbf{R})$) is invariant under $T$ because the derivative of any polynomial of degree at most 4 is also a polynomial with degree at most 4.

*The most famous unsolved problem in functional analysis is called the invariant subspace problem. It deals with invariant subspaces of operators on infinite-dimensional vector spaces.*

Let's look at some easy examples of invariant subspaces. Suppose $T \in \mathcal{L}(V)$. Clearly $\{0\}$ is invariant under $T$. Also, the whole space $V$ is obviously invariant under $T$. Must $T$ have any invariant subspaces other than $\{0\}$ and $V$? Later we will see that this question has an affirmative answer for operators on complex vector spaces with dimension greater than 1 and also for operators on real vector spaces with dimension greater than 2.

If $T \in \mathcal{L}(V)$, then null $T$ is invariant under $T$ (proof: if $u \in$ null $T$, then $Tu = 0$, and hence $Tu \in$ null $T$). Also, range $T$ is invariant under $T$ (proof: if $u \in$ range $T$, then $Tu$ is also in range $T$, by the definition of range). Although null $T$ and range $T$ are invariant under $T$, they do not necessarily provide easy answers to the question about the existence of invariant subspaces other than $\{0\}$ and $V$ because null $T$ may equal $\{0\}$ and range $T$ may equal $V$ (this happens when $T$ is invertible).

We will return later to a deeper study of invariant subspaces. Now we turn to an investigation of the simplest possible nontrivial invariant subspaces—invariant subspaces with dimension 1.

How does an operator behave on an invariant subspace of dimension 1? Subspaces of $V$ of dimension 1 are easy to describe. Take any nonzero vector $u \in V$ and let $U$ equal the set of all scalar multiples of $u$:

**5.2** $$U = \{au : a \in \mathbf{F}\}.$$

Then $U$ is a one-dimensional subspace of $V$, and every one-dimensional subspace of $V$ is of this form. If $u \in V$ and the subspace $U$ defined by 5.2 is invariant under $T \in \mathcal{L}(V)$, then $Tu$ must be in $U$, and hence there must be a scalar $\lambda \in \mathbf{F}$ such that $Tu = \lambda u$. Conversely, if $u$ is a nonzero vector in $V$ such that $Tu = \lambda u$ for some $\lambda \in \mathbf{F}$, then the subspace $U$ defined by 5.2 is a one-dimensional subspace of $V$ invariant under $T$.

The equation

**5.3** $$Tu = \lambda u,$$

which we have just seen is intimately connected with one-dimensional invariant subspaces, is important enough that the vectors $u$ and scalars $\lambda$ satisfying it are given special names. Specifically, a scalar $\lambda \in \mathbf{F}$ is called an ***eigenvalue*** of $T \in \mathcal{L}(V)$ if there exists a nonzero vector $u \in V$ such that $Tu = \lambda u$. We must require $u$ to be nonzero because with $u = 0$ every scalar $\lambda \in \mathbf{F}$ satisfies 5.3. The comments above show that $T$ has a one-dimensional invariant subspace if and only if $T$ has an eigenvalue.

The equation $Tu = \lambda u$ is equivalent to $(T - \lambda I)u = 0$, so $\lambda$ is an eigenvalue of $T$ if and only if $T - \lambda I$ is not injective. By 3.21, $\lambda$ is an eigenvalue of $T$ if and only if $T - \lambda I$ is not invertible, and this happens if and only if $T - \lambda I$ is not surjective.

Suppose $T \in \mathcal{L}(V)$ and $\lambda \in \mathbf{F}$ is an eigenvalue of $T$. A vector $u \in V$ is called an ***eigenvector*** of $T$ (corresponding to $\lambda$) if $Tu = \lambda u$. Because 5.3 is equivalent to $(T - \lambda I)u = 0$, we see that the set of eigenvectors of $T$ corresponding to $\lambda$ equals $\text{null}(T - \lambda I)$. In particular, the set of eigenvectors of $T$ corresponding to $\lambda$ is a subspace of $V$.

*These subspaces are loosely connected to the subject of Herbert Marcuse's well-known book **One-Dimensional Man**.*

*The regrettable word **eigenvalue** is half-German, half-English. The German adjective **eigen** means own in the sense of characterizing some intrinsic property. Some mathematicians use the term **characteristic value** instead of eigenvalue.*

*Some texts define
eigenvectors as we
have, except that 0 is
declared not to be an
eigenvector. With the
definition used here,
the set of eigenvectors
corresponding to a
fixed eigenvalue is a
subspace.*

Let's look at some examples of eigenvalues and eigenvectors. If $a \in \mathbf{F}$, then $aI$ has only one eigenvalue, namely, $a$, and every vector is an eigenvector for this eigenvalue.

For a more complicated example, consider the operator $T \in \mathcal{L}(\mathbf{F}^2)$ defined by

**5.4**                             $T(w, z) = (-z, w).$

If $\mathbf{F} = \mathbf{R}$, then this operator has a nice geometric interpretation: $T$ is just a counterclockwise rotation by $90°$ about the origin in $\mathbf{R}^2$. An operator has an eigenvalue if and only if there exists a nonzero vector in its domain that gets sent by the operator to a scalar multiple of itself. The rotation of a nonzero vector in $\mathbf{R}^2$ obviously never equals a scalar multiple of itself. Conclusion: if $\mathbf{F} = \mathbf{R}$, the operator $T$ defined by 5.4 has no eigenvalues. However, if $\mathbf{F} = \mathbf{C}$, the story changes. To find eigenvalues of $T$, we must find the scalars $\lambda$ such that

$$T(w, z) = \lambda(w, z)$$

has some solution other than $w = z = 0$. For $T$ defined by 5.4, the equation above is equivalent to the simultaneous equations

**5.5**                             $-z = \lambda w, \quad w = \lambda z.$

Substituting the value for $w$ given by the second equation into the first equation gives

$$-z = \lambda^2 z.$$

Now $z$ cannot equal 0 (otherwise 5.5 implies that $w = 0$; we are looking for solutions to 5.5 where $(w, z)$ is not the 0 vector), so the equation above leads to the equation

$$-1 = \lambda^2.$$

The solutions to this equation are $\lambda = i$ or $\lambda = -i$. You should be able to verify easily that $i$ and $-i$ are eigenvalues of $T$. Indeed, the eigenvectors corresponding to the eigenvalue $i$ are the vectors of the form $(w, -wi)$, with $w \in \mathbf{C}$, and the eigenvectors corresponding to the eigenvalue $-i$ are the vectors of the form $(w, wi)$, with $w \in \mathbf{C}$.

Now we show that nonzero eigenvectors corresponding to distinct eigenvalues are linearly independent.

**5.6     Theorem:**  *Let $T \in \mathcal{L}(V)$. Suppose $\lambda_1, \ldots, \lambda_m$ are distinct eigenvalues of $T$ and $v_1, \ldots, v_m$ are corresponding nonzero eigenvectors. Then $(v_1, \ldots, v_m)$ is linearly independent.*

PROOF:   Suppose $(v_1, \ldots, v_m)$ is linearly dependent. Let $k$ be the smallest positive integer such that

**5.7**                           $v_k \in \operatorname{span}(v_1, \ldots, v_{k-1});$

the existence of $k$ with this property follows from the linear dependence lemma (2.4). Thus there exist $a_1, \ldots, a_{k-1} \in \mathbf{F}$ such that

**5.8**                           $v_k = a_1 v_1 + \cdots + a_{k-1} v_{k-1}.$

Apply $T$ to both sides of this equation, getting

$$\lambda_k v_k = a_1 \lambda_1 v_1 + \cdots + a_{k-1} \lambda_{k-1} v_{k-1}.$$

Multiply both sides of 5.8 by $\lambda_k$ and then subtract the equation above, getting

$$0 = a_1(\lambda_k - \lambda_1) v_1 + \cdots + a_{k-1}(\lambda_k - \lambda_{k-1}) v_{k-1}.$$

Because we chose $k$ to be the smallest positive integer satisfying 5.7, $(v_1, \ldots, v_{k-1})$ is linearly independent. Thus the equation above implies that all the $a$'s are 0 (recall that $\lambda_k$ is not equal to any of $\lambda_1, \ldots, \lambda_{k-1}$). However, this means that $v_k$ equals 0 (see 5.8), contradicting our hypothesis that all the $v$'s are nonzero. Therefore our assumption that $(v_1, \ldots, v_m)$ is linearly dependent must have been false.                    ∎

The corollary below states that an operator cannot have more distinct eigenvalues than the dimension of the vector space on which it acts.

**5.9     Corollary:**  *Each operator on $V$ has at most $\dim V$ distinct eigenvalues.*

PROOF:   Let $T \in \mathcal{L}(V)$. Suppose that $\lambda_1, \ldots, \lambda_m$ are distinct eigenvalues of $T$. Let $v_1, \ldots, v_m$ be corresponding nonzero eigenvectors. The last theorem implies that $(v_1, \ldots, v_m)$ is linearly independent. Thus $m \leq \dim V$ (see 2.6), as desired.                    ∎

# *Polynomials Applied to Operators*

The main reason that a richer theory exists for operators (which map a vector space into itself) than for linear maps is that operators can be raised to powers. In this section we define that notion and the key concept of applying a polynomial to an operator.

If $T \in \mathcal{L}(V)$, then $TT$ makes sense and is also in $\mathcal{L}(V)$. We usually write $T^2$ instead of $TT$. More generally, if $m$ is a positive integer, then $T^m$ is defined by

$$T^m = \underbrace{T \ldots T}_{m \text{ times}}.$$

For convenience we define $T^0$ to be the identity operator $I$ on $V$.

Recall from Chapter 3 that if $T$ is an invertible operator, then the inverse of $T$ is denoted by $T^{-1}$. If $m$ is a positive integer, then we define $T^{-m}$ to be $(T^{-1})^m$.

You should verify that if $T$ is an operator, then

$$T^m T^n = T^{m+n} \quad \text{and} \quad (T^m)^n = T^{mn},$$

where $m$ and $n$ are allowed to be arbitrary integers if $T$ is invertible and nonnegative integers if $T$ is not invertible.

If $T \in \mathcal{L}(V)$ and $p \in \mathcal{P}(\mathbf{F})$ is a polynomial given by

$$p(z) = a_0 + a_1 z + a_2 z^2 + \cdots + a_m z^m$$

for $z \in \mathbf{F}$, then $p(T)$ is the operator defined by

$$p(T) = a_0 I + a_1 T + a_2 T^2 + \cdots + a_m T^m.$$

For example, if $p$ is the polynomial defined by $p(z) = z^2$ for $z \in \mathbf{F}$, then $p(T) = T^2$. This is a new use of the symbol $p$ because we are applying it to operators, not just elements of $\mathbf{F}$. If we fix an operator $T \in \mathcal{L}(V)$, then the function from $\mathcal{P}(\mathbf{F})$ to $\mathcal{L}(V)$ given by $p \mapsto p(T)$ is linear, as you should verify.

If $p$ and $q$ are polynomials with coefficients in $\mathbf{F}$, then $pq$ is the polynomial defined by

$$(pq)(z) = p(z)q(z)$$

for $z \in \mathbf{F}$. You should verify that we have the following nice multiplicative property: if $T \in \mathcal{L}(V)$, then

$$(pq)(T) = p(T)q(T)$$

for all polynomials $p$ and $q$ with coefficients in **F**. Note that any two polynomials in $T$ commute, meaning that $p(T)q(T) = q(T)p(T)$, because

$$p(T)q(T) = (pq)(T) = (qp)(T) = q(T)p(T).$$

# Upper-Triangular Matrices

Now we come to one of the central results about operators on complex vector spaces.

**5.10   Theorem:**   *Every operator on a finite-dimensional, nonzero, complex vector space has an eigenvalue.*

PROOF: Suppose $V$ is a complex vector space with dimension $n > 0$ and $T \in \mathcal{L}(V)$. Choose $v \in V$ with $v \neq 0$. Then

$$(v, Tv, T^2v, \ldots, T^nv)$$

cannot be linearly independent because $V$ has dimension $n$ and we have $n + 1$ vectors. Thus there exist complex numbers $a_0, \ldots, a_n$, not all 0, such that

$$0 = a_0v + a_1Tv + \cdots + a_nT^nv.$$

Let $m$ be the largest index such that $a_m \neq 0$. Because $v \neq 0$, the coefficients $a_1, \ldots, a_m$ cannot all be 0, so $0 < m \leq n$. Make the $a$'s the coefficients of a polynomial, which can be written in factored form (see 4.8) as

$$a_0 + a_1z + \cdots + a_nz^n = c(z - \lambda_1)\ldots(z - \lambda_m),$$

where $c$ is a nonzero complex number, each $\lambda_j \in \mathbf{C}$, and the equation holds for all $z \in \mathbf{C}$. We then have

$$\begin{aligned}
0 &= a_0v + a_1Tv + \cdots + a_nT^nv \\
&= (a_0I + a_1T + \cdots + a_nT^n)v \\
&= c(T - \lambda_1I)\ldots(T - \lambda_mI)v,
\end{aligned}$$

which means that $T - \lambda_jI$ is not injective for at least one $j$. In other words, $T$ has an eigenvalue.   ■

*Compare the simple proof of this theorem given here with the standard proof using determinants. With the standard proof, first the difficult concept of determinants must be defined, then an operator with 0 determinant must be shown to be not invertible, then the characteristic polynomial needs to be defined, and by the time the proof of this theorem is reached, no insight remains about why it is true.*

Recall that in Chapter 3 we discussed the matrix of a linear map from one vector space to another vector space. This matrix depended on a choice of a basis for each of the two vector spaces. Now that we are studying operators, which map a vector space to itself, we need only one basis. In addition, now our matrices will be square arrays, rather than the more general rectangular arrays that we considered earlier. Specifically, let $T \in \mathcal{L}(V)$. Suppose $(v_1, \ldots, v_n)$ is a basis of $V$. For each $k = 1, \ldots, n$, we can write

$$Tv_k = a_{1,k}v_1 + \cdots + a_{n,k}v_n,$$

*The $k^{th}$ column of the matrix is formed from the coefficients used to write $Tv_k$ as a linear combination of the $v$'s.*

where $a_{j,k} \in \mathbf{F}$ for $j = 1, \ldots, n$. The $n$-by-$n$ matrix

**5.11**

$$\begin{bmatrix} a_{1,1} & \ldots & a_{1,n} \\ \vdots & & \vdots \\ a_{n,1} & \ldots & a_{n,n} \end{bmatrix}$$

is called the **matrix** of $T$ with respect to the basis $(v_1, \ldots, v_n)$; we denote it by $\mathcal{M}(T, (v_1, \ldots, v_n))$ or just by $\mathcal{M}(T)$ if the basis $(v_1, \ldots, v_n)$ is clear from the context (for example, if only one basis is in sight).

If $T$ is an operator on $\mathbf{F}^n$ and no basis is specified, you should assume that the basis in question is the standard one (where the $j^{th}$ basis vector is 1 in the $j^{th}$ slot and 0 in all the other slots). You can then think of the $j^{th}$ column of $\mathcal{M}(T)$ as $T$ applied to the $j^{th}$ basis vector.

A central goal of linear algebra is to show that given an operator $T \in \mathcal{L}(V)$, there exists a basis of $V$ with respect to which $T$ has a reasonably simple matrix. To make this vague formulation ("reasonably simple" is not precise language) a bit more concrete, we might try to make $\mathcal{M}(T)$ have many 0's.

If $V$ is a complex vector space, then we already know enough to show that there is a basis of $V$ with respect to which the matrix of $T$ has 0's everywhere in the first column, except possibly the first entry. In other words, there is a basis of $V$ with respect to which the matrix of $T$ looks like

*We often use $*$ to denote matrix entries that we do not know about or that are irrelevant to the questions being discussed.*

$$\begin{bmatrix} \lambda & & \\ 0 & & * \\ \vdots & & \\ 0 & & \end{bmatrix};$$

here the $*$ denotes the entries in all the columns other than the first column. To prove this, let $\lambda$ be an eigenvalue of $T$ (one exists by 5.10) and let $v$ be a corresponding nonzero eigenvector. Extend $(v)$ to a basis of $V$. Then the matrix of $T$ with respect to this basis has the form above. Soon we will see that we can choose a basis of $V$ with respect to which the matrix of $T$ has even more 0's.

The **diagonal** of a square matrix consists of the entries along the straight line from the upper left corner to the bottom right corner. For example, the diagonal of the matrix 5.11 consists of the entries $a_{1,1}, a_{2,2}, \ldots, a_{n,n}$.

A matrix is called **upper triangular** if all the entries below the diagonal equal 0. For example, the 4-by-4 matrix

$$\begin{bmatrix} 6 & 2 & 7 & 5 \\ 0 & 6 & 1 & 3 \\ 0 & 0 & 7 & 9 \\ 0 & 0 & 0 & 8 \end{bmatrix}$$

is upper triangular. Typically we represent an upper-triangular matrix in the form

$$\begin{bmatrix} \lambda_1 & & * \\ & \ddots & \\ 0 & & \lambda_n \end{bmatrix};$$

the 0 in the matrix above indicates that all entries below the diagonal in this $n$-by-$n$ matrix equal 0. Upper-triangular matrices can be considered reasonably simple—for $n$ large, an $n$-by-$n$ upper-triangular matrix has almost half its entries equal to 0.

The following proposition demonstrates a useful connection between upper-triangular matrices and invariant subspaces.

**5.12   Proposition:**   *Suppose $T \in \mathcal{L}(V)$ and $(v_1, \ldots, v_n)$ is a basis of $V$. Then the following are equivalent:*

(a)     *the matrix of $T$ with respect to $(v_1, \ldots, v_n)$ is upper triangular;*

(b)     $Tv_k \in \mathrm{span}(v_1, \ldots, v_k)$ *for each $k = 1, \ldots, n$;*

(c)     $\mathrm{span}(v_1, \ldots, v_k)$ *is invariant under $T$ for each $k = 1, \ldots, n$.*

PROOF:   The equivalence of (a) and (b) follows easily from the definitions and a moment's thought. Obviously (c) implies (b). Thus to complete the proof, we need only prove that (b) implies (c). So suppose that (b) holds. Fix $k \in \{1, \ldots, n\}$. From (b), we know that

$$Tv_1 \in \text{span}(v_1) \subset \text{span}(v_1, \ldots, v_k);$$
$$Tv_2 \in \text{span}(v_1, v_2) \subset \text{span}(v_1, \ldots, v_k);$$
$$\vdots$$
$$Tv_k \in \text{span}(v_1, \ldots, v_k).$$

Thus if $v$ is a linear combination of $(v_1, \ldots, v_k)$, then

$$Tv \in \text{span}(v_1, \ldots, v_k).$$

In other words, $\text{span}(v_1, \ldots, v_k)$ is invariant under $T$, completing the proof. ∎

Now we can show that for each operator on a complex vector space, there is a basis of the vector space with respect to which the matrix of the operator has only 0's below the diagonal. In Chapter 8 we will improve even this result.

*This theorem does not hold on real vector spaces because the first vector in a basis with respect to which an operator has an upper-triangular matrix must be an eigenvector of the operator. Thus if an operator on a real vector space has no eigenvalues (we have seen an example on $\mathbf{R}^2$), then there is no basis with respect to which the operator has an upper-triangular matrix.*

**5.13   Theorem:** *Suppose $V$ is a complex vector space and $T \in \mathcal{L}(V)$. Then $T$ has an upper-triangular matrix with respect to some basis of $V$.*

PROOF:   We will use induction on the dimension of $V$. Clearly the desired result holds if $\dim V = 1$.

Suppose now that $\dim V > 1$ and the desired result holds for all complex vector spaces whose dimension is less than the dimension of $V$. Let $\lambda$ be any eigenvalue of $T$ (5.10 guarantees that $T$ has an eigenvalue). Let

$$U = \text{range}(T - \lambda I).$$

Because $T - \lambda I$ is not surjective (see 3.21), $\dim U < \dim V$. Furthermore, $U$ is invariant under $T$. To prove this, suppose $u \in U$. Then

$$Tu = (T - \lambda I)u + \lambda u.$$

Obviously $(T - \lambda I)u \in U$ (from the definition of $U$) and $\lambda u \in U$. Thus the equation above shows that $Tu \in U$. Hence $U$ is invariant under $T$, as claimed.

Thus $T|_U$ is an operator on $U$. By our induction hypothesis, there is a basis $(u_1, \ldots, u_m)$ of $U$ with respect to which $T|_U$ has an upper-triangular matrix. Thus for each $j$ we have (using 5.12)

**5.14**         $$Tu_j = (T|_U)(u_j) \in \text{span}(u_1, \ldots, u_j).$$

Extend $(u_1, \ldots, u_m)$ to a basis $(u_1, \ldots, u_m, v_1, \ldots, v_n)$ of $V$. For each $k$, we have
$$Tv_k = (T - \lambda I)v_k + \lambda v_k.$$

The definition of $U$ shows that $(T - \lambda I)v_k \in U = \mathrm{span}(u_1, \ldots, u_m)$. Thus the equation above shows that

**5.15**                  $Tv_k \in \mathrm{span}(u_1, \ldots, u_m, v_1, \ldots, v_k).$

From 5.14 and 5.15, we conclude (using 5.12) that $T$ has an upper-triangular matrix with respect to the basis $(u_1, \ldots, u_m, v_1, \ldots, v_n)$. ∎

How does one determine from looking at the matrix of an operator whether the operator is invertible? If we are fortunate enough to have a basis with respect to which the matrix of the operator is upper triangular, then this problem becomes easy, as the following proposition shows.

**5.16 Proposition:** *Suppose $T \in \mathcal{L}(V)$ has an upper-triangular matrix with respect to some basis of $V$. Then $T$ is invertible if and only if all the entries on the diagonal of that upper-triangular matrix are nonzero.*

PROOF: Suppose $(v_1, \ldots, v_n)$ is a basis of $V$ with respect to which $T$ has an upper-triangular matrix

**5.17**        $\mathcal{M}(T, (v_1, \ldots, v_n)) = \begin{bmatrix} \lambda_1 & & & * \\ & \lambda_2 & & \\ & & \ddots & \\ 0 & & & \lambda_n \end{bmatrix}.$

We need to prove that $T$ is not invertible if and only if one of the $\lambda_k$'s equals 0.

First we will prove that if one of the $\lambda_k$'s equals 0, then $T$ is not invertible. If $\lambda_1 = 0$, then $Tv_1 = 0$ (from 5.17) and hence $T$ is not invertible, as desired. So suppose that $1 < k \le n$ and $\lambda_k = 0$. Then, as can be seen from 5.17, $T$ maps each of the vectors $v_1, \ldots, v_{k-1}$ into $\mathrm{span}(v_1, \ldots, v_{k-1})$. Because $\lambda_k = 0$, the matrix representation 5.17 also implies that $Tv_k \in \mathrm{span}(v_1, \ldots, v_{k-1})$. Thus we can define a linear map

$$S : \mathrm{span}(v_1, \ldots, v_k) \to \mathrm{span}(v_1, \ldots, v_{k-1})$$

by $Sv = Tv$ for $v \in \mathrm{span}(v_1, \ldots, v_k)$. In other words, $S$ is just $T$ restricted to $\mathrm{span}(v_1, \ldots, v_k)$.

Note that $\mathrm{span}(v_1, \ldots, v_k)$ has dimension $k$ and $\mathrm{span}(v_1, \ldots, v_{k-1})$ has dimension $k - 1$ (because $(v_1, \ldots, v_n)$ is linearly independent). Because $\mathrm{span}(v_1, \ldots, v_k)$ has a larger dimension than $\mathrm{span}(v_1, \ldots, v_{k-1})$, no linear map from $\mathrm{span}(v_1, \ldots, v_k)$ to $\mathrm{span}(v_1, \ldots, v_{k-1})$ is injective (see 3.5). Thus there exists a nonzero vector $v \in \mathrm{span}(v_1, \ldots, v_k)$ such that $Sv = 0$. Hence $Tv = 0$, and thus $T$ is not invertible, as desired.

To prove the other direction, now suppose that $T$ is not invertible. Thus $T$ is not injective (see 3.21), and hence there exists a nonzero vector $v \in V$ such that $Tv = 0$. Because $(v_1, \ldots, v_n)$ is a basis of $V$, we can write

$$v = a_1 v_1 + \cdots + a_k v_k,$$

where $a_1, \ldots, a_k \in \mathbf{F}$ and $a_k \neq 0$ (represent $v$ as a linear combination of $(v_1, \ldots, v_n)$ and then choose $k$ to be the largest index with a nonzero coefficient). Thus

$$
\begin{aligned}
0 &= Tv \\
0 &= T(a_1 v_1 + \cdots + a_k v_k) \\
&= (a_1 T v_1 + \cdots + a_{k-1} T v_{k-1}) + a_k T v_k.
\end{aligned}
$$

The last term in parentheses is in $\mathrm{span}(v_1, \ldots, v_{k-1})$ (because of the upper-triangular form of 5.17). Thus the last equation shows that $a_k T v_k \in \mathrm{span}(v_1, \ldots, v_{k-1})$. Multiplying by $1/a_k$, which is allowed because $a_k \neq 0$, we conclude that $T v_k \in \mathrm{span}(v_1, \ldots, v_{k-1})$. Thus when $T v_k$ is written as a linear combination of the basis $(v_1, \ldots, v_n)$, the coefficient of $v_k$ will be 0. In other words, $\lambda_k$ in 5.17 must be 0, completing the proof. ∎

*Powerful numeric techniques exist for finding good approximations to the eigenvalues of an operator from its matrix.*

Unfortunately no method exists for exactly computing the eigenvalues of a typical operator from its matrix (with respect to an arbitrary basis). However, if we are fortunate enough to find a basis with respect to which the matrix of the operator is upper triangular, then the problem of computing the eigenvalues becomes trivial, as the following proposition shows.

**5.18 Proposition:** *Suppose $T \in \mathcal{L}(V)$ has an upper-triangular matrix with respect to some basis of $V$. Then the eigenvalues of $T$ consist precisely of the entries on the diagonal of that upper-triangular matrix.*

PROOF: Suppose $(v_1, \ldots, v_n)$ is a basis of $V$ with respect to which $T$ has an upper-triangular matrix

$$\mathcal{M}(T, (v_1, \ldots, v_n)) = \begin{bmatrix} \lambda_1 & & & * \\ & \lambda_2 & & \\ & & \ddots & \\ 0 & & & \lambda_n \end{bmatrix}.$$

Let $\lambda \in \mathbf{F}$. Then

$$\mathcal{M}(T - \lambda I, (v_1, \ldots, v_n)) = \begin{bmatrix} \lambda_1 - \lambda & & & * \\ & \lambda_2 - \lambda & & \\ & & \ddots & \\ 0 & & & \lambda_n - \lambda \end{bmatrix}.$$

Hence $T - \lambda I$ is not invertible if and only if $\lambda$ equals one of the $\lambda_j'$s (see 5.16). In other words, $\lambda$ is an eigenvalue of $T$ if and only if $\lambda$ equals one of the $\lambda_j'$s, as desired. ∎

## Diagonal Matrices

A **diagonal matrix** is a square matrix that is 0 everywhere except possibly along the diagonal. For example,

$$\begin{bmatrix} 8 & 0 & 0 \\ 0 & 2 & 0 \\ 0 & 0 & 5 \end{bmatrix}$$

is a diagonal matrix. Obviously every diagonal matrix is upper triangular, although in general a diagonal matrix has many more 0's than an upper-triangular matrix.

An operator $T \in \mathcal{L}(V)$ has a diagonal matrix

$$\begin{bmatrix} \lambda_1 & & 0 \\ & \ddots & \\ 0 & & \lambda_n \end{bmatrix}$$

with respect to a basis $(v_1, \ldots, v_n)$ of $V$ if and only

$$Tv_1 = \lambda_1 v_1$$

$$\vdots$$

$$Tv_n = \lambda_n v_n;$$

this follows immediately from the definition of the matrix of an operator with respect to a basis. Thus an operator $T \in \mathcal{L}(V)$ has a diagonal matrix with respect to some basis of $V$ if and only if $V$ has a basis consisting of eigenvectors of $T$.

If an operator has a diagonal matrix with respect to some basis, then the entries along the diagonal are precisely the eigenvalues of the operator; this follows from 5.18 (or you may want to find an easier proof that works only for diagonal matrices).

Unfortunately not every operator has a diagonal matrix with respect to some basis. This sad state of affairs can arise even on complex vector spaces. For example, consider $T \in \mathcal{L}(\mathbf{C}^2)$ defined by

**5.19**                                    $T(w, z) = (z, 0)$.

As you should verify, 0 is the only eigenvalue of this operator and the corresponding set of eigenvectors is the one-dimensional subspace $\{(w, 0) \in \mathbf{C}^2 : w \in \mathbf{C}\}$. Thus there are not enough linearly independent eigenvectors of $T$ to form a basis of the two-dimensional space $\mathbf{C}^2$. Hence $T$ does not have a diagonal matrix with respect to any basis of $\mathbf{C}^2$.

The next proposition shows that if an operator has as many distinct eigenvalues as the dimension of its domain, then the operator has a diagonal matrix with respect to some operator. However, some operators with fewer eigenvalues also have diagonal matrices (in other words, the converse of the next proposition is not true). For example, the operator $T$ defined on the three-dimensional space $\mathbf{F}^3$ by

$$T(z_1, z_2, z_3) = (4z_1, 4z_2, 5z_3)$$

has only two eigenvalues (4 and 5), but this operator has a diagonal matrix with respect to the standard basis.

*Later we will find other conditions that imply that certain operators have a diagonal matrix with respect to some basis (see 7.9 and 7.13).*

**5.20   Proposition:** *If $T \in \mathcal{L}(V)$ has* $\dim V$ *distinct eigenvalues, then $T$ has a diagonal matrix with respect to some basis of $V$.*

PROOF:   Suppose that $T \in \mathcal{L}(V)$ has $\dim V$ distinct eigenvalues $\lambda_1, \ldots, \lambda_{\dim V}$. For each $j$, let $v_j \in V$ be a nonzero eigenvector corresponding to the eigenvalue $\lambda_j$. Because nonzero eigenvectors corresponding to distinct eigenvalues are linearly independent (see 5.6), $(v_1, \ldots, v_{\dim V})$ is linearly independent. A linearly independent list of

$\dim V$ vectors in $V$ is a basis of $V$ (see 2.17); thus $(v_1, \ldots, v_{\dim V})$ is a basis of $V$. With respect to this basis consisting of eigenvectors, $T$ has a diagonal matrix. ∎

We close this section with a proposition giving several conditions on an operator that are equivalent to its having a diagonal matrix with respect to some basis.

**5.21   Proposition:** *Suppose $T \in \mathcal{L}(V)$. Let $\lambda_1, \ldots, \lambda_m$ denote the distinct eigenvalues of $T$. Then the following are equivalent:*

*For complex vector spaces, we will extend this list of equivalences later (see Exercises 16 and 23 in Chapter 8).*

(a)   *$T$ has a diagonal matrix with respect to some basis of $V$;*

(b)   *$V$ has a basis consisting of eigenvectors of $T$;*

(c)   *there exist one-dimensional subspaces $U_1, \ldots, U_n$ of $V$, each invariant under $T$, such that*

$$V = U_1 \oplus \cdots \oplus U_n;$$

(d)   $V = \mathrm{null}(T - \lambda_1 I) \oplus \cdots \oplus \mathrm{null}(T - \lambda_m I);$

(e)   $\dim V = \dim \mathrm{null}(T - \lambda_1 I) + \cdots + \dim \mathrm{null}(T - \lambda_m I).$

PROOF:   We have already shown that (a) and (b) are equivalent.

Suppose that (b) holds; thus $V$ has a basis $(v_1, \ldots, v_n)$ consisting of eigenvectors of $T$. For each $j$, let $U_j = \mathrm{span}(v_j)$. Obviously each $U_j$ is a one-dimensional subspace of $V$ that is invariant under $T$ (because each $v_j$ is an eigenvector of $T$). Because $(v_1, \ldots, v_n)$ is a basis of $V$, each vector in $V$ can be written uniquely as a linear combination of $(v_1, \ldots, v_n)$. In other words, each vector in $V$ can be written uniquely as a sum $u_1 + \cdots + u_n$, where each $u_j \in U_j$. Thus $V = U_1 \oplus \cdots \oplus U_n$. Hence (b) implies (c).

Suppose now that (c) holds; thus there are one-dimensional subspaces $U_1, \ldots, U_n$ of $V$, each invariant under $T$, such that

$$V = U_1 \oplus \cdots \oplus U_n.$$

For each $j$, let $v_j$ be a nonzero vector in $U_j$. Then each $v_j$ is an eigenvector of $T$. Because each vector in $V$ can be written uniquely as a sum $u_1 + \cdots + u_n$, where each $u_j \in U_j$ (so each $u_j$ is a scalar multiple of $v_j$), we see that $(v_1, \ldots, v_n)$ is a basis of $V$. Thus (c) implies (b).

LIVERPOOL JOHN MOORES UNIVERSITY
LEARNING SERVICES

At this stage of the proof we know that (a), (b), and (c) are all equivalent. We will finish the proof by showing that (b) implies (d), that (d) implies (e), and that (e) implies (b).

Suppose that (b) holds; thus $V$ has a basis consisting of eigenvectors of $T$. Thus every vector in $V$ is a linear combination of eigenvectors of $T$. Hence

**5.22**              $V = \text{null}(T - \lambda_1 I) + \cdots + \text{null}(T - \lambda_m I).$

To show that the sum above is a direct sum, suppose that

$$0 = u_1 + \cdots + u_m,$$

where each $u_j \in \text{null}(T - \lambda_j I)$. Because nonzero eigenvectors corresponding to distinct eigenvalues are linearly independent, this implies (apply 5.6 to the sum of the nonzero vectors on the right side of the equation above) that each $u_j$ equals 0. This implies (using 1.8) that the sum in 5.22 is a direct sum, completing the proof that (b) implies (d).

That (d) implies (e) follows immediately from Exercise 17 in Chapter 2.

Finally, suppose that (e) holds; thus

**5.23**         $\dim V = \dim \text{null}(T - \lambda_1 I) + \cdots + \dim \text{null}(T - \lambda_m I).$

Choose a basis of each $\text{null}(T - \lambda_j I)$; put all these bases together to form a list $(v_1, \ldots, v_n)$ of eigenvectors of $T$, where $n = \dim V$ (by 5.23). To show that this list is linearly independent, suppose

$$a_1 v_1 + \cdots + a_n v_n = 0,$$

where $a_1, \ldots, a_n \in \mathbf{F}$. For each $j = 1, \ldots, m$, let $u_j$ denote the sum of all the terms $a_k v_k$ such that $v_k \in \text{null}(T - \lambda_j I)$. Thus each $u_j$ is an eigenvector of $T$ with eigenvalue $\lambda_j$, and

$$u_1 + \cdots + u_m = 0.$$

Because nonzero eigenvectors corresponding to distinct eigenvalues are linearly independent, this implies (apply 5.6 to the sum of the nonzero vectors on the left side of the equation above) that each $u_j$ equals 0. Because each $u_j$ is a sum of terms $a_k v_k$, where the $v_k$'s were chosen to be a basis of $\text{null}(T - \lambda_j I)$, this implies that all the $a_k$'s equal 0. Thus $(v_1, \ldots, v_n)$ is linearly independent and hence is a basis of $V$ (by 2.17). Thus (e) implies (b), completing the proof.  ∎

# Invariant Subspaces on Real Vector Spaces

We know that every operator on a complex vector space has an eigen-value (see 5.10 for the precise statement). We have also seen an example showing that the analogous statement is false on real vector spaces. In other words, an operator on a nonzero real vector space may have no invariant subspaces of dimension 1. However, we now show that an invariant subspace of dimension 1 or 2 always exists.

**5.24  Theorem:** *Every operator on a finite-dimensional, nonzero, real vector space has an invariant subspace of dimension 1 or 2.*

PROOF: Suppose $V$ is a real vector space with dimension $n > 0$ and $T \in \mathcal{L}(V)$. Choose $v \in V$ with $v \neq 0$. Then

$$(v, Tv, T^2v, \ldots, T^nv)$$

cannot be linearly independent because $V$ has dimension $n$ and we have $n + 1$ vectors. Thus there exist real numbers $a_0, \ldots, a_n$, not all 0, such that

$$0 = a_0v + a_1Tv + \cdots + a_nT^nv.$$

Make the $a$'s the coefficients of a polynomial, which can be written in factored form (see 4.14) as

$$a_0 + a_1x + \cdots + a_nx^n$$
$$= c(x - \lambda_1)\ldots(x - \lambda_m)(x^2 + \alpha_1x + \beta_1)\ldots(x^2 + \alpha_Mx + \beta_M),$$

*Here either $m$ or $M$ might equal 0.*

where $c$ is a nonzero real number, each $\lambda_j$, $\alpha_j$, and $\beta_j$ is real, $m+M \geq 1$, and the equation holds for all $x \in \mathbf{R}$. We then have

$$0 = a_0v + a_1Tv + \cdots + a_nT^nv$$
$$= (a_0I + a_1T + \cdots + a_nT^n)v$$
$$= c(T - \lambda_1I)\ldots(T - \lambda_mI)(T^2 + \alpha_1T + \beta_1I)\ldots(T^2 + \alpha_MT + \beta_MI)v,$$

which means that $T - \lambda_jI$ is not injective for at least one $j$ or that $(T^2 + \alpha_jT + \beta_jI)$ is not injective for at least one $j$. If $T - \lambda_jI$ is not injective for at least one $j$, then $T$ has an eigenvalue and hence a one-dimensional invariant subspace. Let's consider the other possibility. In other words, suppose that $(T^2 + \alpha_jT + \beta_jI)$ is not injective for some $j$. Thus there exists a nonzero vector $u \in V$ such that

**5.25**                     $$T^2u + \alpha_j Tu + \beta_j u = 0.$$

We will complete the proof by showing that span$(u, Tu)$, which clearly has dimension 1 or 2, is invariant under $T$. To do this, consider a typical element of span$(u, Tu)$ of the form $au + bTu$, where $a, b \in \mathbf{R}$. Then

$$T(au + bTu) = aTu + bT^2u$$
$$= aTu - b\alpha_j Tu - b\beta_j u,$$

where the last equality comes from solving for $T^2u$ in 5.25. The equation above shows that $T(au + bTu) \in$ span$(u, Tu)$. Thus span$(u, Tu)$ is invariant under $T$, as desired. ∎

We will need one new piece of notation for the next proof. Suppose $U$ and $W$ are subspaces of $V$ with

$$V = U \oplus W.$$

Each vector $v \in V$ can be written uniquely in the form

$$v = u + w,$$

*$P_{U,W}$ is often called the **projection** onto U with null space W.*

where $u \in U$ and $w \in W$. With this representation, define $P_{U,W} \in \mathcal{L}(V)$ by

$$P_{U,W}v = u.$$

You should verify that $P_{U,W}v = v$ if and only if $v \in U$. Interchanging the roles of $U$ and $W$ in the representation above, we have $P_{W,U}v = w$. Thus $v = P_{U,W}v + P_{W,U}v$ for every $v \in V$. You should verify that $P_{U,W}{}^2 = P_{U,W}$; furthermore range $P_{U,W} = U$ and null $P_{U,W} = W$.

We have seen an example of an operator on $\mathbf{R}^2$ with no eigenvalues. The following theorem shows that no such example exists on $\mathbf{R}^3$.

**5.26 Theorem:** *Every operator on an odd-dimensional real vector space has an eigenvalue.*

PROOF: Suppose $V$ is a real vector space with odd dimension. We will prove that every operator on $V$ has an eigenvalue by induction (in steps of size 2) on the dimension of $V$. To get started, note that the desired result obviously holds if dim $V = 1$.

Now suppose that dim $V$ is an odd number greater than 1. Using induction, we can assume that the desired result holds for all operators on all real vector spaces with dimension 2 less than $\dim V$. Suppose $T \in \mathcal{L}(V)$. We need to prove that $T$ has an eigenvalue. If it does, we are done. If not, then by 5.24 there is a two-dimensional subspace $U$ of $V$ that is invariant under $T$. Let $W$ be any subspace of $V$ such that

$$V = U \oplus W;$$

2.13 guarantees that such a $W$ exists.

Because $W$ has dimension 2 less than $\dim V$, we would like to apply our induction hypothesis to $T|_W$. However, $W$ might not be invariant under $T$, meaning that $T|_W$ might not be an operator on $W$. We will compose with the projection $P_{W,U}$ to get an operator on $W$. Specifically, define $S \in \mathcal{L}(W)$ by

$$Sw = P_{W,U}(Tw)$$

for $w \in W$. By our induction hypothesis, $S$ has an eigenvalue $\lambda$. We will show that this $\lambda$ is also an eigenvalue for $T$.

Let $w \in W$ be a nonzero eigenvector for $S$ corresponding to the eigenvalue $\lambda$; thus $(S - \lambda I)w = 0$. We would be done if $w$ were an eigenvector for $T$ with eigenvalue $\lambda$; unfortunately that need not be true. So we will look for an eigenvector of $T$ in $U + \operatorname{span}(w)$. To do that, consider a typical vector $u + aw$ in $U + \operatorname{span}(w)$, where $u \in U$ and $a \in \mathbf{R}$. We have

$$
\begin{aligned}
(T - \lambda I)(u + aw) &= Tu - \lambda u + a(Tw - \lambda w) \\
&= Tu - \lambda u + a(P_{U,W}(Tw) + P_{W,U}(Tw) - \lambda w) \\
&= Tu - \lambda u + a(P_{U,W}(Tw) + Sw - \lambda w) \\
&= Tu - \lambda u + aP_{U,W}(Tw).
\end{aligned}
$$

Note that on the right side of the last equation, $Tu \in U$ (because $U$ is invariant under $T$), $\lambda u \in U$ (because $u \in U$), and $aP_{U,W}(Tw) \in U$ (from the definition of $P_{U,W}$). Thus $T - \lambda I$ maps $U + \operatorname{span}(w)$ into $U$. Because $U + \operatorname{span}(w)$ has a larger dimension than $U$, this means that $(T - \lambda I)|_{U+\operatorname{span}(w)}$ is not injective (see 3.5). In other words, there exists a nonzero vector $v \in U + \operatorname{span}(w) \subset V$ such that $(T - \lambda I)v = 0$. Thus $T$ has an eigenvalue, as desired.                                        ∎

## *Exercises*

1.  Suppose $T \in \mathcal{L}(V)$. Prove that if $U_1, \ldots, U_m$ are subspaces of $V$ invariant under $T$, then $U_1 + \cdots + U_m$ is invariant under $T$.

2.  Suppose $T \in \mathcal{L}(V)$. Prove that the intersection of any collection of subspaces of $V$ invariant under $T$ is invariant under $T$.

3.  Prove or give a counterexample: if $U$ is a subspace of $V$ that is invariant under every operator on $V$, then $U = \{0\}$ or $U = V$.

4.  Suppose that $S, T \in \mathcal{L}(V)$ are such that $ST = TS$. Prove that $\text{null}(T - \lambda I)$ is invariant under $S$ for every $\lambda \in \mathbf{F}$.

5.  Define $T \in \mathcal{L}(\mathbf{F}^2)$ by

    $$T(w, z) = (z, w).$$

    Find all eigenvalues and eigenvectors of $T$.

6.  Define $T \in \mathcal{L}(\mathbf{F}^3)$ by

    $$T(z_1, z_2, z_3) = (2z_2, 0, 5z_3).$$

    Find all eigenvalues and eigenvectors of $T$.

7.  Suppose $n$ is a positive integer and $T \in \mathcal{L}(\mathbf{F}^n)$ is defined by

    $$T(x_1, \ldots, x_n) = (x_1 + \cdots + x_n, \ldots, x_1 + \cdots + x_n);$$

    in other words, $T$ is the operator whose matrix (with respect to the standard basis) consists of all 1's. Find all eigenvalues and eigenvectors of $T$.

8.  Find all eigenvalues and eigenvectors of the backward shift operator $T \in \mathcal{L}(\mathbf{F}^\infty)$ defined by

    $$T(z_1, z_2, z_3, \ldots) = (z_2, z_3, \ldots).$$

9.  Suppose $T \in \mathcal{L}(V)$ and $\dim \text{range}\, T = k$. Prove that $T$ has at most $k + 1$ distinct eigenvalues.

10. Suppose $T \in \mathcal{L}(V)$ is invertible and $\lambda \in \mathbf{F} \setminus \{0\}$. Prove that $\lambda$ is an eigenvalue of $T$ if and only if $\frac{1}{\lambda}$ is an eigenvalue of $T^{-1}$.

11.   Suppose $S, T \in \mathcal{L}(V)$. Prove that $ST$ and $TS$ have the same eigen-values.

12.   Suppose $T \in \mathcal{L}(V)$ is such that every vector in $V$ is an eigenvector of $T$. Prove that $T$ is a scalar multiple of the identity operator.

13.   Suppose $T \in \mathcal{L}(V)$ is such that every subspace of $V$ with dimension $\dim V - 1$ is invariant under $T$. Prove that $T$ is a scalar multiple of the identity operator.

14.   Suppose $S, T \in \mathcal{L}(V)$ and $S$ is invertible. Prove that if $p \in \mathcal{P}(\mathbf{F})$ is a polynomial, then

$$p(STS^{-1}) = Sp(T)S^{-1}.$$

15.   Suppose $\mathbf{F} = \mathbf{C}$, $T \in \mathcal{L}(V)$, $p \in \mathcal{P}(\mathbf{C})$, and $a \in \mathbf{C}$. Prove that $a$ is an eigenvalue of $p(T)$ if and only if $a = p(\lambda)$ for some eigenvalue $\lambda$ of $T$.

16.   Show that the result in the previous exercise does not hold if $\mathbf{C}$ is replaced with $\mathbf{R}$.

17.   Suppose $V$ is a complex vector space and $T \in \mathcal{L}(V)$. Prove that $T$ has an invariant subspace of dimension $j$ for each $j = 1, \ldots, \dim V$.

18.   Give an example of an operator whose matrix with respect to some basis contains only 0's on the diagonal, but the operator is invertible.

*These two exercises show that 5.16 fails without the hypothesis that an upper-triangular matrix is under consideration.*

19.   Give an example of an operator whose matrix with respect to some basis contains only nonzero numbers on the diagonal, but the operator is not invertible.

20.   Suppose that $T \in \mathcal{L}(V)$ has $\dim V$ distinct eigenvalues and that $S \in \mathcal{L}(V)$ has the same eigenvectors as $T$ (not necessarily with the same eigenvalues). Prove that $ST = TS$.

21.   Suppose $P \in \mathcal{L}(V)$ and $P^2 = P$. Prove that $V = \operatorname{null} P \oplus \operatorname{range} P$.

22.   Suppose $V = U \oplus W$, where $U$ and $W$ are nonzero subspaces of $V$. Find all eigenvalues and eigenvectors of $P_{U,W}$.

LIVERPOOL
JOHN MOORES UNIVERSITY
AVRIL ROBARTS LRC
TEL. 0151 231 4022

23.    Give an example of an operator $T \in \mathcal{L}(\mathbf{R}^4)$ such that $T$ has no (real) eigenvalues.

24.    Suppose $V$ is a real vector space and $T \in \mathcal{L}(V)$ has no eigenvalues. Prove that every subspace of $V$ invariant under $T$ has even dimension.

# CHAPTER 6

# *Inner-Product Spaces*

In making the definition of a vector space, we generalized the linear structure (addition and scalar multiplication) of $\mathbf{R}^2$ and $\mathbf{R}^3$. We ignored other important features, such as the notions of length and angle. These ideas are embedded in the concept we now investigate, inner products.

Recall that $\mathbf{F}$ denotes $\mathbf{R}$ or $\mathbf{C}$.
Also, $V$ is a finite-dimensional, nonzero vector space over $\mathbf{F}$.

# Inner Products

To motivate the concept of inner product, let's think of vectors in $\mathbf{R}^2$ and $\mathbf{R}^3$ as arrows with initial point at the origin. The length of a vector $x$ in $\mathbf{R}^2$ or $\mathbf{R}^3$ is called the ***norm*** of $x$, denoted $\|x\|$. Thus for $x = (x_1, x_2) \in \mathbf{R}^2$, we have $\|x\| = \sqrt{x_1{}^2 + x_2{}^2}$.

*If we think of vectors as points instead of arrows, then $\|x\|$ should be interpreted as the distance from the point $x$ to the origin.*

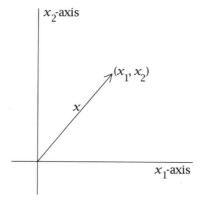

*The length of this vector $x$ is $\sqrt{x_1{}^2 + x_2{}^2}$.*

Similarly, for $x = (x_1, x_2, x_3) \in \mathbf{R}^3$, we have $\|x\| = \sqrt{x_1{}^2 + x_2{}^2 + x_3{}^2}$. Even though we cannot draw pictures in higher dimensions, the generalization to $\mathbf{R}^n$ is obvious: we define the norm of $x = (x_1, \ldots, x_n) \in \mathbf{R}^n$ by

$$\|x\| = \sqrt{x_1{}^2 + \cdots + x_n{}^2}.$$

The norm is not linear on $\mathbf{R}^n$. To inject linearity into the discussion, we introduce the dot product. For $x, y \in \mathbf{R}^n$, the ***dot product*** of $x$ and $y$, denoted $x \cdot y$, is defined by

$$x \cdot y = x_1 y_1 + \cdots + x_n y_n,$$

where $x = (x_1, \ldots, x_n)$ and $y = (y_1, \ldots, y_n)$. Note that the dot product of two vectors in $\mathbf{R}^n$ is a number, not a vector. Obviously $x \cdot x = \|x\|^2$ for all $x \in \mathbf{R}^n$. In particular, $x \cdot x \geq 0$ for all $x \in \mathbf{R}^n$, with equality if and only if $x = 0$. Also, if $y \in \mathbf{R}^n$ is fixed, then clearly the map from $\mathbf{R}^n$ to $\mathbf{R}$ that sends $x \in \mathbf{R}^n$ to $x \cdot y$ is linear. Furthermore, $x \cdot y = y \cdot x$ for all $x, y \in \mathbf{R}^n$.

An inner product is a generalization of the dot product. At this point you should be tempted to guess that an inner product is defined by abstracting the properties of the dot product discussed in the paragraph above. For real vector spaces, that guess is correct. However, so that we can make a definition that will be useful for both real and complex vector spaces, we need to examine the complex case before making the definition.

Recall that if $\lambda = a + bi$, where $a, b \in \mathbf{R}$, then the absolute value of $\lambda$ is defined by

$$|\lambda| = \sqrt{a^2 + b^2},$$

the complex conjugate of $\lambda$ is defined by

$$\bar{\lambda} = a - bi,$$

and the equation

$$|\lambda|^2 = \lambda\bar{\lambda}$$

connects these two concepts (see page 69 for the definitions and the basic properties of the absolute value and complex conjugate). For $z = (z_1, \ldots, z_n) \in \mathbf{C}^n$, we define the norm of $z$ by

$$\|z\| = \sqrt{|z_1|^2 + \cdots + |z_n|^2}.$$

The absolute values are needed because we want $\|z\|$ to be a nonnegative number. Note that

$$\|z\|^2 = z_1\overline{z_1} + \cdots + z_n\overline{z_n}.$$

We want to think of $\|z\|^2$ as the inner product of $z$ with itself, as we did in $\mathbf{R}^n$. The equation above thus suggests that the inner product of $w = (w_1, \ldots, w_n) \in \mathbf{C}^n$ with $z$ should equal

$$w_1\overline{z_1} + \cdots + w_n\overline{z_n}.$$

If the roles of the $w$ and $z$ were interchanged, the expression above would be replaced with its complex conjugate. In other words, we should expect that the inner product of $w$ with $z$ equals the complex conjugate of the inner product of $z$ with $w$. With that motivation, we are now ready to define an inner product on $V$, which may be a real or a complex vector space.

An ***inner product*** on $V$ is a function that takes each ordered pair $(u, v)$ of elements of $V$ to a number $\langle u, v \rangle \in \mathbf{F}$ and has the following properties:

*If z is a complex*
*number, then the*
*statement z ≥ 0 means*
*that z is real and*
*nonnegative.*

**positivity**
   $\langle v, v \rangle \geq 0$ for all $v \in V$;

**definiteness**
   $\langle v, v \rangle = 0$ if and only if $v = 0$;

**additivity in first slot**
   $\langle u + v, w \rangle = \langle u, w \rangle + \langle v, w \rangle$ for all $u, v, w \in V$;

**homogeneity in first slot**
   $\langle av, w \rangle = a \langle v, w \rangle$ for all $a \in \mathbf{F}$ and all $v, w \in V$;

**conjugate symmetry**
   $\langle v, w \rangle = \overline{\langle w, v \rangle}$ for all $v, w \in V$.

Recall that every real number equals its complex conjugate. Thus if we are dealing with a real vector space, then in the last condition above we can dispense with the complex conjugate and simply state that $\langle v, w \rangle = \langle w, v \rangle$ for all $v, w \in V$.

An ***inner-product space*** is a vector space $V$ along with an inner product on $V$.

The most important example of an inner-product space is $\mathbf{F}^n$. We can define an inner product on $\mathbf{F}^n$ by

*If we are dealing with*
$\mathbf{R}^n$ *rather than* $\mathbf{C}^n$*, then*
*again the complex*
*conjugate can be*
*ignored.*

**6.1**          $\langle (w_1, \ldots, w_n), (z_1, \ldots, z_n) \rangle = w_1 \overline{z_1} + \cdots + w_n \overline{z_n},$

as you should verify. This inner product, which provided our motivation for the definition of an inner product, is called the ***Euclidean inner product*** on $\mathbf{F}^n$. When $\mathbf{F}^n$ is referred to as an inner-product space, you should assume that the inner product is the Euclidean inner product unless explicitly told otherwise.

There are other inner products on $\mathbf{F}^n$ in addition to the Euclidean inner product. For example, if $c_1, \ldots, c_n$ are positive numbers, then we can define an inner product on $\mathbf{F}^n$ by

$$\langle (w_1, \ldots, w_n), (z_1, \ldots, z_n) \rangle = c_1 w_1 \overline{z_1} + \cdots + c_n w_n \overline{z_n},$$

as you should verify. Of course, if all the $c$'s equal 1, then we get the Euclidean inner product.

As another example of an inner-product space, consider the vector space $\mathcal{P}_m(\mathbf{F})$ of all polynomials with coefficients in $\mathbf{F}$ and degree at most $m$. We can define an inner product on $\mathcal{P}_m(\mathbf{F})$ by

**6.2**                     $$\langle p,q \rangle = \int_0^1 p(x)\overline{q(x)}\,dx,$$

as you should verify. Once again, if $\mathbf{F} = \mathbf{R}$, then the complex conjugate
is not needed.

> Let's agree for the rest of this chapter that
> $V$ is a finite-dimensional inner-product space over $\mathbf{F}$.

In the definition of an inner product, the conditions of additivity
and homogeneity in the first slot can be combined into a requirement
of linearity in the first slot. More precisely, for each fixed $w \in V$, the
function that takes $v$ to $\langle v, w \rangle$ is a linear map from $V$ to $\mathbf{F}$. Because
every linear map takes 0 to 0, we must have

$$\langle 0, w \rangle = 0$$

for every $w \in V$. Thus we also have

$$\langle w, 0 \rangle = 0$$

for every $w \in V$ (by the conjugate symmetry property).

In an inner-product space, we have additivity in the second slot as
well as the first slot. Proof:

$$\begin{aligned}
\langle u, v + w \rangle &= \overline{\langle v + w, u \rangle} \\
&= \overline{\langle v, u \rangle + \langle w, u \rangle} \\
&= \overline{\langle v, u \rangle} + \overline{\langle w, u \rangle} \\
&= \langle u, v \rangle + \langle u, w \rangle;
\end{aligned}$$

here $u, v, w \in V$.

In an inner-product space, we have conjugate homogeneity in the
second slot, meaning that $\langle u, av \rangle = \bar{a}\langle u, v \rangle$ for all scalars $a \in \mathbf{F}$.
Proof:

$$\begin{aligned}
\langle u, av \rangle &= \overline{\langle av, u \rangle} \\
&= \overline{a\langle v, u \rangle} \\
&= \bar{a}\overline{\langle v, u \rangle} \\
&= \bar{a}\langle u, v \rangle;
\end{aligned}$$

here $a \in \mathbf{F}$ and $u, v \in V$. Note that in a real vector space, conjugate
homogeneity is the same as homogeneity.

# Norms

For $v \in V$, we define the **norm** of $v$, denoted $\|v\|$, by

$$\|v\| = \sqrt{\langle v, v \rangle}.$$

For example, if $(z_1, \ldots, z_n) \in \mathbf{F}^n$ (with the Euclidean inner product), then

$$\|(z_1, \ldots, z_n)\| = \sqrt{|z_1|^2 + \cdots + |z_n|^2}.$$

As another example, if $p \in \mathcal{P}_m(\mathbf{F})$ (with inner product given by 6.2), then

$$\|p\| = \sqrt{\int_0^1 |p(x)|^2 \, dx}.$$

Note that $\|v\| = 0$ if and only if $v = 0$ (because $\langle v, v \rangle = 0$ if and only if $v = 0$). Another easy property of the norm is that $\|av\| = |a| \, \|v\|$ for all $a \in \mathbf{F}$ and all $v \in V$. Here's the proof:

$$\begin{aligned}
\|av\|^2 &= \langle av, av \rangle \\
&= a \langle v, av \rangle \\
&= a\bar{a} \langle v, v \rangle \\
&= |a|^2 \|v\|^2;
\end{aligned}$$

taking square roots now gives the desired equality. This proof illustrates a general principle: working with norms squared is usually easier than working directly with norms.

*Some mathematicians use the term **perpendicular**, which means the same as orthogonal.*

Two vectors $u, v \in V$ are said to be **orthogonal** if $\langle u, v \rangle = 0$. Note that the order of the vectors does not matter because $\langle u, v \rangle = 0$ if and only if $\langle v, u \rangle = 0$. Instead of saying that $u$ and $v$ are orthogonal, sometimes we say that $u$ is orthogonal to $v$. Clearly 0 is orthogonal to every vector. Furthermore, 0 is the only vector that is orthogonal to itself.

*The word **orthogonal** comes from the Greek word **orthogonios**, which means right-angled.*

For the special case where $V = \mathbf{R}^2$, the next theorem is over 2,500 years old.

**6.3**   **Pythagorean Theorem:** *If $u, v$ are orthogonal vectors in $V$, then*

**6.4**
$$\|u + v\|^2 = \|u\|^2 + \|v\|^2.$$

PROOF:  Suppose that $u, v$ are orthogonal vectors in $V$. Then

$$\|u + v\|^2 = \langle u + v, u + v \rangle$$
$$= \|u\|^2 + \|v\|^2 + \langle u, v \rangle + \langle v, u \rangle$$
$$= \|u\|^2 + \|v\|^2,$$

as desired.                                                                                   ∎

Suppose $u, v \in V$. We would like to write $u$ as a scalar multiple of $v$ plus a vector $w$ orthogonal to $v$, as suggested in the next picture.

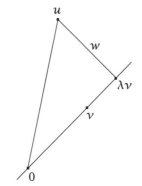

An orthogonal decomposition

To discover how to write $u$ as a scalar multiple of $v$ plus a vector orthogonal to $v$, let $a \in \mathbf{F}$ denote a scalar. Then

$$u = av + (u - av).$$

Thus we need to choose $a$ so that $v$ is orthogonal to $(u - av)$. In other words, we want

$$0 = \langle u - av, v \rangle = \langle u, v \rangle - a\|v\|^2.$$

The equation above shows that we should choose $a$ to be $\langle u, v \rangle / \|v\|^2$ (assume that $v \neq 0$ to avoid division by 0). Making this choice of $a$, we can write

**6.5**             $$u = \frac{\langle u, v \rangle}{\|v\|^2} v + \left( u - \frac{\langle u, v \rangle}{\|v\|^2} v \right).$$

As you should verify, if $v \neq 0$ then the equation above writes $u$ as a scalar multiple of $v$ plus a vector orthogonal to $v$.

The equation above will be used in the proof of the next theorem, which gives one of the most important inequalities in mathematics.

*The proof of the Pythagorean theorem shows that 6.4 holds if and only if $\langle u, v \rangle + \langle v, u \rangle$, which equals $2\operatorname{Re}\langle u, v \rangle$, is 0. Thus the converse of the Pythagorean theorem holds in real inner-product spaces.*

*In 1821 the French
mathematician
Augustin-Louis Cauchy
showed that this
inequality holds for the
inner product defined
by 6.1. In 1886 the
German mathematician
Herman Schwarz
showed that this
inequality holds for the
inner product defined
by 6.2.*

**6.6**   **Cauchy-Schwarz Inequality:** *If $u, v \in V$, then*

**6.7**
$$|\langle u, v \rangle| \le \|u\| \, \|v\|.$$

*This inequality is an equality if and only if one of $u, v$ is a scalar multiple of the other.*

PROOF:   Let $u, v \in V$. If $v = 0$, then both sides of 6.7 equal 0 and the desired inequality holds. Thus we can assume that $v \ne 0$. Consider the orthogonal decomposition

$$u = \frac{\langle u, v \rangle}{\|v\|^2} v + w,$$

where $w$ is orthogonal to $v$ (here $w$ equals the second term on the right side of 6.5). By the Pythagorean theorem,

$$\|u\|^2 = \left\| \frac{\langle u, v \rangle}{\|v\|^2} v \right\|^2 + \|w\|^2$$

$$= \frac{|\langle u, v \rangle|^2}{\|v\|^2} + \|w\|^2$$

**6.8**
$$\ge \frac{|\langle u, v \rangle|^2}{\|v\|^2}.$$

Multiplying both sides of this inequality by $\|v\|^2$ and then taking square roots gives the Cauchy-Schwarz inequality 6.7.

Looking at the proof of the Cauchy-Schwarz inequality, note that 6.7 is an equality if and only if 6.8 is an equality. Obviously this happens if and only if $w = 0$. But $w = 0$ if and only if $u$ is a multiple of $v$ (see 6.5). Thus the Cauchy-Schwarz inequality is an equality if and only if $u$ is a scalar multiple of $v$ or $v$ is a scalar multiple of $u$ (or both; the phrasing has been chosen to cover cases in which either $u$ or $v$ equals 0).   ∎

The next result is called the triangle inequality because of its geometric interpretation that the length of any side of a triangle is less than the sum of the lengths of the other two sides.

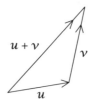

*The triangle inequality*

**6.9**    **Triangle Inequality:** *If* $u, v \in V$, *then*

*The triangle inequality can be used to show that the shortest path between two points is a straight line segment.*

**6.10**                         $$\|u + v\| \leq \|u\| + \|v\|.$$

*This inequality is an equality if and only if one of* $u, v$ *is a nonnegative multiple of the other.*

PROOF:   Let $u, v \in V$. Then

$$
\begin{aligned}
\|u + v\|^2 &= \langle u + v, u + v \rangle \\
&= \langle u, u \rangle + \langle v, v \rangle + \langle u, v \rangle + \langle v, u \rangle \\
&= \langle u, u \rangle + \langle v, v \rangle + \langle u, v \rangle + \overline{\langle u, v \rangle} \\
&= \|u\|^2 + \|v\|^2 + 2\,\mathrm{Re}\langle u, v \rangle
\end{aligned}
$$

**6.11**                  $$\leq \|u\|^2 + \|v\|^2 + 2|\langle u, v \rangle|$$

**6.12**                  $$\leq \|u\|^2 + \|v\|^2 + 2\|u\|\,\|v\|$$

$$= (\|u\| + \|v\|)^2,$$

where 6.12 follows from the Cauchy-Schwarz inequality (6.6). Taking square roots of both sides of the inequality above gives the triangle inequality 6.10.

The proof above shows that the triangle inequality 6.10 is an equality if and only if we have equality in 6.11 and 6.12. Thus we have equality in the triangle inequality 6.10 if and only if

**6.13**                         $$\langle u, v \rangle = \|u\|\|v\|.$$

If one of $u, v$ is a nonnegative multiple of the other, then 6.13 holds, as you should verify. Conversely, suppose 6.13 holds. Then the condition for equality in the Cauchy-Schwarz inequality (6.6) implies that one of $u, v$ must be a scalar multiple of the other. Clearly 6.13 forces the scalar in question to be nonnegative, as desired.                    ∎

The next result is called the parallelogram equality because of its geometric interpretation: in any parallelogram, the sum of the squares of the lengths of the diagonals equals the sum of the squares of the lengths of the four sides.

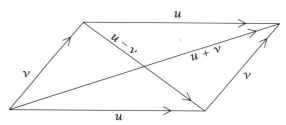

*The parallelogram equality*

**6.14   Parallelogram Equality:**  *If* $u, v \in V$, *then*

$$\|u + v\|^2 + \|u - v\|^2 = 2(\|u\|^2 + \|v\|^2).$$

PROOF:  Let $u, v \in V$. Then

$$
\begin{aligned}
\|u + v\|^2 + \|u - v\|^2 &= \langle u + v, u + v \rangle + \langle u - v, u - v \rangle \\
&= \|u\|^2 + \|v\|^2 + \langle u, v \rangle + \langle v, u \rangle \\
&\quad + \|u\|^2 + \|v\|^2 - \langle u, v \rangle - \langle v, u \rangle \\
&= 2(\|u\|^2 + \|v\|^2),
\end{aligned}
$$

as desired.                                                                      ∎

## *Orthonormal Bases*

A list of vectors is called ***orthonormal*** if the vectors in it are pairwise orthogonal and each vector has norm 1. In other words, a list $(e_1, \ldots, e_m)$ of vectors in $V$ is orthonormal if $\langle e_j, e_k \rangle$ equals 0 when $j \neq k$ and equals 1 when $j = k$ (for $j, k = 1, \ldots, m$). For example, the standard basis in $\mathbf{F}^n$ is orthonormal. Orthonormal lists are particularly easy to work with, as illustrated by the next proposition.

**6.15   Proposition:**  *If* $(e_1, \ldots, e_m)$ *is an orthonormal list of vectors in $V$, then*

$$\|a_1 e_1 + \cdots + a_m e_m\|^2 = |a_1|^2 + \cdots + |a_m|^2$$

*for all* $a_1, \ldots, a_m \in \mathbf{F}$.

PROOF:  Because each $e_j$ has norm 1, this follows easily from repeated applications of the Pythagorean theorem (6.3).                        ∎

Now we have the following easy but important corollary.

**6.16   Corollary:**   *Every orthonormal list of vectors is linearly independent.*

PROOF:   Suppose $(e_1, \ldots, e_m)$ is an orthonormal list of vectors in $V$ and $a_1, \ldots, a_m \in \mathbf{F}$ are such that

$$a_1 e_1 + \cdots + a_m e_m = 0.$$

Then $|a_1|^2 + \cdots + |a_m|^2 = 0$ (by 6.15), which means that all the $a_j$'s are 0, as desired.                                                                                        ∎

An **orthonormal basis** of $V$ is an orthonormal list of vectors in $V$ that is also a basis of $V$. For example, the standard basis is an orthonormal basis of $\mathbf{F}^n$. Every orthonormal list of vectors in $V$ with length $\dim V$ is automatically an orthonormal basis of $V$ (proof: by the previous corollary, any such list must be linearly independent; because it has the right length, it must be a basis—see 2.17). To illustrate this principle, consider the following list of four vectors in $\mathbf{R}^4$:

$$((\tfrac{1}{2}, \tfrac{1}{2}, \tfrac{1}{2}, \tfrac{1}{2}), (\tfrac{1}{2}, \tfrac{1}{2}, -\tfrac{1}{2}, -\tfrac{1}{2}), (\tfrac{1}{2}, -\tfrac{1}{2}, -\tfrac{1}{2}, \tfrac{1}{2}), (-\tfrac{1}{2}, \tfrac{1}{2}, -\tfrac{1}{2}, \tfrac{1}{2})).$$

The verification that this list is orthonormal is easy (do it!); because we have an orthonormal list of length four in a four-dimensional vector space, it must be an orthonormal basis.

In general, given a basis $(e_1, \ldots, e_n)$ of $V$ and a vector $v \in V$, we know that there is some choice of scalars $a_1, \ldots, a_m$ such that

$$v = a_1 e_1 + \cdots + a_n e_n,$$

but finding the $a_j$'s can be difficult. The next theorem shows, however, that this is easy for an orthonormal basis.

**6.17   Theorem:**   *Suppose $(e_1, \ldots, e_n)$ is an orthonormal basis of $V$. Then*

*The importance of orthonormal bases stems mainly from this theorem.*

**6.18**                        $$v = \langle v, e_1 \rangle e_1 + \cdots + \langle v, e_n \rangle e_n$$

*and*

**6.19**                        $$\|v\|^2 = |\langle v, e_1 \rangle|^2 + \cdots + |\langle v, e_n \rangle|^2$$

*for every $v \in V$.*

PROOF:  Let $v \in V$. Because $(e_1, \ldots, e_n)$ is a basis of $V$, there exist scalars $a_1, \ldots, a_n$ such that

$$v = a_1 e_1 + \cdots + a_n e_n.$$

Take the inner product of both sides of this equation with $e_j$, getting $\langle v, e_j \rangle = a_j$. Thus 6.18 holds. Clearly 6.19 follows from 6.18 and 6.15.  ∎

Now that we understand the usefulness of orthonormal bases, how do we go about finding them? For example, does $\mathcal{P}_m(\mathbf{F})$, with inner product given by integration on $[0, 1]$ (see 6.2), have an orthonormal basis? As we will see, the next result will lead to answers to these questions. The algorithm used in the next proof is called the ***Gram-Schmidt procedure***. It gives a method for turning a linearly independent list into an orthonormal list with the same span as the original list.

*The Dutch mathematician Jorgen Gram (1850-1916) and the German mathematician Erhard Schmidt (1876-1959) popularized this algorithm for constructing orthonormal lists.*

**6.20   Gram-Schmidt:**   *If $(v_1, \ldots, v_m)$ is a linearly independent list of vectors in V, then there exists an orthonormal list $(e_1, \ldots, e_m)$ of vectors in V such that*

**6.21**                    $\mathrm{span}(v_1, \ldots, v_j) = \mathrm{span}(e_1, \ldots, e_j)$

*for $j = 1, \ldots, m$.*

PROOF:  Suppose $(v_1, \ldots, v_m)$ is a linearly independent list of vectors in $V$. To construct the $e$'s, start by setting $e_1 = v_1 / \|v_1\|$. This satisfies 6.21 for $j = 1$. We will choose $e_2, \ldots, e_m$ inductively, as follows. Suppose $j > 1$ and an orthornormal list $(e_1, \ldots, e_{j-1})$ has been chosen so that

**6.22**                    $\mathrm{span}(v_1, \ldots, v_{j-1}) = \mathrm{span}(e_1, \ldots, e_{j-1}).$

Let

**6.23**        $e_j = \dfrac{v_j - \langle v_j, e_1 \rangle e_1 - \cdots - \langle v_j, e_{j-1} \rangle e_{j-1}}{\|v_j - \langle v_j, e_1 \rangle e_1 - \cdots - \langle v_j, e_{j-1} \rangle e_{j-1}\|}.$

Note that $v_j \notin \mathrm{span}(v_1, \ldots, v_{j-1})$ (because $(v_1, \ldots, v_m)$ is linearly independent) and thus $v_j \notin \mathrm{span}(e_1, \ldots, e_{j-1})$. Hence we are not dividing by 0 in the equation above, and so $e_j$ is well defined. Dividing a vector by its norm produces a new vector with norm 1; thus $\|e_j\| = 1$.

Let $1 \le k < j$. Then

$$\langle e_j, e_k \rangle = \left\langle \frac{v_j - \langle v_j, e_1 \rangle e_1 - \cdots - \langle v_j, e_{j-1} \rangle e_{j-1}}{\| v_j - \langle v_j, e_1 \rangle e_1 - \cdots - \langle v_j, e_{j-1} \rangle e_{j-1} \|}, e_k \right\rangle$$

$$= \frac{\langle v_j, e_k \rangle - \langle v_j, e_k \rangle}{\| v_j - \langle v_j, e_1 \rangle e_1 - \cdots - \langle v_j, e_{j-1} \rangle e_{j-1} \|}$$

$$= 0.$$

Thus $(e_1, \ldots, e_j)$ is an orthonormal list.

From 6.23, we see that $v_j \in \mathrm{span}(e_1, \ldots, e_j)$. Combining this information with 6.22 shows that

$$\mathrm{span}(v_1, \ldots, v_j) \subset \mathrm{span}(e_1, \ldots, e_j).$$

Both lists above are linearly independent (the $v$'s by hypothesis, the $e$'s by orthonormality and 6.16). Thus both subspaces above have dimension $j$, and hence they must be equal, completing the proof.  ∎

Now we can settle the question of the existence of orthonormal bases.

**6.24  Corollary:** *Every finite-dimensional inner-product space has an orthonormal basis.*

PROOF:  Choose a basis of $V$. Apply the Gram-Schmidt procedure (6.20) to it, producing an orthonormal list. This orthonormal list is linearly independent (by 6.16) and its span equals $V$. Thus it is an orthonormal basis of $V$.  ∎

*Until this corollary, nothing we had done with inner-product spaces required our standing assumption that V is finite dimensional.*

As we will soon see, sometimes we need to know not only that an orthonormal basis exists, but also that any orthonormal list can be extended to an orthonormal basis. In the next corollary, the Gram-Schmidt procedure shows that such an extension is always possible.

**6.25  Corollary:** *Every orthonormal list of vectors in V can be extended to an orthonormal basis of V.*

PROOF:  Suppose $(e_1, \ldots, e_m)$ is an orthonormal list of vectors in $V$. Then $(e_1, \ldots, e_m)$ is linearly independent (by 6.16), and hence it can be extended to a basis $(e_1, \ldots, e_m, v_1, \ldots, v_n)$ of $V$ (see 2.12). Now apply the Gram-Schmidt procedure (6.20) to $(e_1, \ldots, e_m, v_1, \ldots, v_n)$, producing an orthonormal list

**6.26**                          $(e_1, \ldots, e_m, f_1, \ldots, f_n);$

here the Gram-Schmidt procedure leaves the first $m$ vectors unchanged because they are already orthonormal. Clearly 6.26 is an orthonormal basis of $V$ because it is linearly independent (by 6.16) and its span equals $V$. Hence we have our extension of $(e_1, \ldots, e_m)$ to an orthonormal basis of $V$.                                                      ∎

Recall that a matrix is called upper triangular if all entries below the diagonal equal 0. In other words, an upper-triangular matrix looks like this:

$$\begin{bmatrix} * & & * \\ & \ddots & \\ 0 & & * \end{bmatrix}.$$

In the last chapter we showed that if $V$ is a complex vector space, then for each operator on $V$ there is a basis with respect to which the matrix of the operator is upper triangular (see 5.13). Now that we are dealing with inner-product spaces, we would like to know when there exists an *orthonormal* basis with respect to which we have an upper-triangular matrix. The next corollary shows that the existence of any basis with respect to which $T$ has an upper-triangular matrix implies the existence of an orthonormal basis with this property. This result is true on both real and complex vector spaces (though on a real vector space, the hypothesis holds only for some operators).

**6.27    Corollary:**  *Suppose $T \in \mathcal{L}(V)$. If $T$ has an upper-triangular matrix with respect to some basis of $V$, then $T$ has an upper-triangular matrix with respect to some orthonormal basis of $V$.*

PROOF:  Suppose $T$ has an upper-triangular matrix with respect to some basis $(v_1, \ldots, v_n)$ of $V$. Thus $\text{span}(v_1, \ldots, v_j)$ is invariant under $T$ for each $j = 1, \ldots, n$ (see 5.12).

Apply the Gram-Schmidt procedure to $(v_1, \ldots, v_n)$, producing an orthonormal basis $(e_1, \ldots, e_n)$ of $V$. Because

$$\text{span}(e_1, \ldots, e_j) = \text{span}(v_1, \ldots, v_j)$$

for each $j$ (see 6.21), we conclude that $\operatorname{span}(e_1, \ldots, e_j)$ is invariant under $T$ for each $j = 1, \ldots, n$. Thus, by 5.12, $T$ has an upper-triangular matrix with respect to the orthonormal basis $(e_1, \ldots, e_n)$. ∎

The next result is an important application of the corollary above.

**6.28 Corollary:** *Suppose $V$ is a complex vector space and $T \in \mathcal{L}(V)$. Then $T$ has an upper-triangular matrix with respect to some orthonormal basis of $V$.*

PROOF: This follows immediately from 5.13 and 6.27. ∎

*This result is sometimes called Schur's theorem. The German mathematician Issai Schur published the first proof of this result in 1909.*

# Orthogonal Projections and Minimization Problems

If $U$ is a subset of $V$, then the **orthogonal complement** of $U$, denoted $U^{\perp}$, is the set of all vectors in $V$ that are orthogonal to every vector in $U$:

$$U^{\perp} = \{v \in V : \langle v, u \rangle = 0 \text{ for all } u \in U\}.$$

You should verify that $U^{\perp}$ is always a subspace of $V$, that $V^{\perp} = \{0\}$, and that $\{0\}^{\perp} = V$. Also note that if $U_1 \subset U_2$, then $U_1^{\perp} \supset U_2^{\perp}$.

Recall that if $U_1, U_2$ are subspaces of $V$, then $V$ is the direct sum of $U_1$ and $U_2$ (written $V = U_1 \oplus U_2$) if each element of $V$ can be written in exactly one way as a vector in $U_1$ plus a vector in $U_2$. The next theorem shows that every subspace of an inner-product space leads to a natural direct sum decomposition of the whole space.

**6.29 Theorem:** *If $U$ is a subspace of $V$, then*

$$V = U \oplus U^{\perp}.$$

PROOF: Suppose that $U$ is a subspace of $V$. First we will show that

**6.30** $$V = U + U^{\perp}.$$

To do this, suppose $v \in V$. Let $(e_1, \ldots, e_m)$ be an orthonormal basis of $U$. Obviously

**6.31**
$$v = \underbrace{\langle v, e_1 \rangle e_1 + \cdots + \langle v, e_m \rangle e_m}_{u} + \underbrace{v - \langle v, e_1 \rangle e_1 - \cdots - \langle v, e_m \rangle e_m}_{w}.$$

Clearly $u \in U$. Because $(e_1, \ldots, e_m)$ is an orthonormal list, for each $j$ we have

$$\langle w, e_j \rangle = \langle v, e_j \rangle - \langle v, e_j \rangle$$
$$= 0.$$

Thus $w$ is orthogonal to every vector in $\operatorname{span}(e_1, \ldots, e_m)$. In other words, $w \in U^\perp$. Thus we have written $v = u + w$, where $u \in U$ and $w \in U^\perp$, completing the proof of 6.30.

If $v \in U \cap U^\perp$, then $v$ (which is in $U$) is orthogonal to every vector in $U$ (including $v$ itself), which implies that $\langle v, v \rangle = 0$, which implies that $v = 0$. Thus

**6.32**                                    $$U \cap U^\perp = \{0\}.$$

Now 6.30 and 6.32 imply that $V = U \oplus U^\perp$ (see 1.9).    ∎

The next corollary is an important consequence of the last theorem.

**6.33    Corollary:**  *If $U$ is a subspace of $V$, then*

$$U = (U^\perp)^\perp.$$

PROOF:  Suppose that $U$ is a subspace of $V$. First we will show that

**6.34**                                    $$U \subset (U^\perp)^\perp.$$

To do this, suppose that $u \in U$. Then $\langle u, v \rangle = 0$ for every $v \in U^\perp$ (by the definition of $U^\perp$). Because $u$ is orthogonal to every vector in $U^\perp$, we have $u \in (U^\perp)^\perp$, completing the proof of 6.34.

To prove the inclusion in the other direction, suppose $v \in (U^\perp)^\perp$. By 6.29, we can write $v = u + w$, where $u \in U$ and $w \in U^\perp$. We have $v - u = w \in U^\perp$. Because $v \in (U^\perp)^\perp$ and $u \in (U^\perp)^\perp$ (from 6.34), we have $v - u \in (U^\perp)^\perp$. Thus $v - u \in U^\perp \cap (U^\perp)^\perp$, which implies that $v - u$ is orthogonal to itself, which implies that $v - u = 0$, which implies that $v = u$, which implies that $v \in U$. Thus $(U^\perp)^\perp \subset U$, which along with 6.34 completes the proof.    ∎

Suppose $U$ is a subspace of $V$. The decomposition $V = U \oplus U^\perp$ given by 6.29 means that each vector $v \in V$ can be written uniquely in the form

$$v = u + w,$$

where $u \in U$ and $w \in U^\perp$. We use this decomposition to define an operator on $V$, denoted $P_U$, called the **orthogonal projection** of $V$ onto $U$. For $v \in V$, we define $P_U v$ to be the vector $u$ in the decomposition above. In the notation introduced in the last chapter, we have $P_U = P_{U,U^\perp}$. You should verify that $P_U \in \mathcal{L}(V)$ and that it has the following properties:

- range $P_U = U$;

- null $P_U = U^\perp$;

- $v - P_U v \in U^\perp$ for every $v \in V$;

- $P_U{}^2 = P_U$;

- $\|P_U v\| \leq \|v\|$ for every $v \in V$.

Furthermore, from the decomposition 6.31 used in the proof of 6.29 we see that if $(e_1, \ldots, e_m)$ is an orthonormal basis of $U$, then

**6.35** $$P_U v = \langle v, e_1 \rangle e_1 + \cdots + \langle v, e_m \rangle e_m$$

for every $v \in V$.

The following problem often arises: given a subspace $U$ of $V$ and a point $v \in V$, find a point $u \in U$ such that $\|v - u\|$ is as small as possible. The next proposition shows that this minimization problem is solved by taking $u = P_U v$.

**6.36 Proposition:** *Suppose $U$ is a subspace of $V$ and $v \in V$. Then*

$$\|v - P_U v\| \leq \|v - u\|$$

*for every $u \in U$. Furthermore, if $u \in U$ and the inequality above is an equality, then $u = P_U v$.*

*The remarkable simplicity of the solution to this minimization problem has led to many applications of inner-product spaces outside of pure mathematics.*

PROOF: Suppose $u \in U$. Then

**6.37** $$\|v - P_U v\|^2 \leq \|v - P_U v\|^2 + \|P_U v - u\|^2$$
**6.38** $$= \|(v - P_U v) + (P_U v - u)\|^2$$
$$= \|v - u\|^2,$$

where 6.38 comes from the Pythagorean theorem (6.3), which applies because $v - P_U v \in U^\perp$ and $P_U v - u \in U$. Taking square roots gives the desired inequality.

Our inequality is an equality if and only if 6.37 is an equality, which happens if and only if $\|P_U v - u\| = 0$, which happens if and only if $u = P_U v$.                                                                ∎

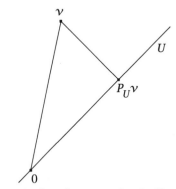

*$P_U v$ is the closest point in $U$ to $v$.*

The last proposition is often combined with the formula 6.35 to compute explicit solutions to minimization problems. As an illustration of this procedure, consider the problem of finding a polynomial $u$ with real coefficients and degree at most 5 that on the interval $[-\pi, \pi]$ approximates $\sin x$ as well as possible, in the sense that

$$\int_{-\pi}^{\pi} |\sin x - u(x)|^2 \, dx$$

is as small as possible. To solve this problem, let $C[-\pi, \pi]$ denote the real vector space of continuous real-valued functions on $[-\pi, \pi]$ with inner product

**6.39**                    $$\langle f, g \rangle = \int_{-\pi}^{\pi} f(x)g(x) \, dx.$$

Let $v \in C[-\pi, \pi]$ be the function defined by $v(x) = \sin x$. Let $U$ denote the subspace of $C[-\pi, \pi]$ consisting of the polynomials with real coefficients and degree at most 5. Our problem can now be re-formulated as follows: find $u \in U$ such that $\|v - u\|$ is as small as possible.

To compute the solution to our approximation problem, first apply the Gram-Schmidt procedure (using the inner product given by 6.39)

to the basis $(1, x, x^2, x^3, x^4, x^5)$ of $U$, producing an orthonormal basis $(e_1, e_2, e_3, e_4, e_5, e_6)$ of $U$. Then, again using the inner product given by 6.39, compute $P_U v$ using 6.35 (with $m = 6$). Doing this computation shows that $P_U v$ is the function

*A machine that can perform integrations is useful here.*

**6.40**          $0.987862x - 0.155271x^3 + 0.00564312x^5,$

where the $\pi$'s that appear in the exact answer have been replaced with a good decimal approximation.

By 6.36, the polynomial above should be about as good an approximation to $\sin x$ on $[-\pi, \pi]$ as is possible using polynomials of degree at most 5. To see how good this approximation is, the picture below shows the graphs of both $\sin x$ and our approximation 6.40 over the interval $[-\pi, \pi]$.

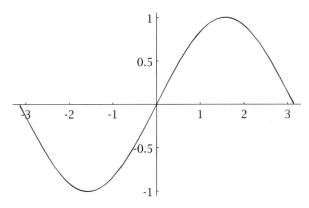

*Graphs of* $\sin x$ *and its approximation 6.40*

Our approximation 6.40 is so accurate that the two graphs are almost identical—our eyes may see only one graph!

Another well-known approximation to $\sin x$ by a polynomial of degree 5 is given by the Taylor polynomial

**6.41**                    $$x - \frac{x^3}{3!} + \frac{x^5}{5!}.$$

To see how good this approximation is, the next picture shows the graphs of both $\sin x$ and the Taylor polynomial 6.41 over the interval $[-\pi, \pi]$.

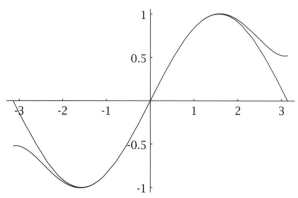

*Graphs of* $\sin x$ *and the Taylor polynomial 6.41*

The Taylor polynomial is an excellent approximation to $\sin x$ for $x$ near 0. But the picture above shows that for $|x| > 2$, the Taylor polynomial is not so accurate, especially compared to 6.40. For example, taking $x = 3$, our approximation 6.40 estimates $\sin 3$ with an error of about 0.001, but the Taylor series 6.41 estimates $\sin 3$ with an error of about 0.4. Thus at $x = 3$, the error in the Taylor series is hundreds of times larger than the error given by 6.40. Linear algebra has helped us discover an approximation to $\sin x$ that improves upon what we learned in calculus!

We derived our approximation 6.40 by using 6.35 and 6.36. Our standing assumption that $V$ is finite dimensional fails when $V$ equals $C[-\pi, \pi]$, so we need to justify our use of those results in this case. First, reread the proof of 6.29, which states that if $U$ is a subspace of $V$, then

**6.42**                    $$V = U \oplus U^{\perp}.$$

*If we allow V to be infinite dimensional and allow U to be an infinite-dimensional subspace of V, then 6.42 is not necessarily true without additional hypotheses.*

Note that the proof uses the finite dimensionality of $U$ (to get a basis of $U$) but that it works fine regardless of whether or not $V$ is finite dimensional. Second, note that the definition and properties of $P_U$ (including 6.35) require only 6.29 and thus require only that $U$ (but not necessarily $V$) be finite dimensional. Finally, note that the proof of 6.36 does not require the finite dimensionality of $V$. Conclusion: for $v \in V$ and $U$ a subspace of $V$, the procedure discussed above for finding the vector $u \in U$ that makes $\|v - u\|$ as small as possible works if $U$ is finite dimensional, regardless of whether or not $V$ is finite dimensional. In the example above $U$ was indeed finite dimensional (we had $\dim U = 6$), so everything works as expected.

# Linear Functionals and Adjoints

A **linear functional** on $V$ is a linear map from $V$ to the scalars $\mathbf{F}$. For example, the function $\varphi \colon \mathbf{F}^3 \to \mathbf{F}$ defined by

**6.43**
$$\varphi(z_1, z_2, z_3) = 2z_1 - 5z_2 + z_3$$

is a linear functional on $\mathbf{F}^3$. As another example, consider the inner-product space $\mathcal{P}_6(\mathbf{R})$ (here the inner product is multiplication followed by integration on $[0,1]$; see 6.2). The function $\varphi \colon \mathcal{P}_6(\mathbf{R}) \to \mathbf{R}$ defined by

**6.44**
$$\varphi(p) = \int_0^1 p(x)(\cos x)\, dx$$

is a linear functional on $\mathcal{P}_6(\mathbf{R})$.

If $v \in V$, then the map that sends $u$ to $\langle u, v \rangle$ is a linear functional on $V$. The next result shows that every linear functional on $V$ is of this form. To illustrate this theorem, note that for the linear functional $\varphi$ defined by 6.43, we can take $v = (2, -5, 1) \in \mathbf{F}^3$. The linear functional $\varphi$ defined by 6.44 better illustrates the power of the theorem below because for this linear functional, there is no obvious candidate for $v$ (the function $\cos x$ is not eligible because it is not an element of $\mathcal{P}_6(\mathbf{R})$).

**6.45    Theorem:** *Suppose $\varphi$ is a linear functional on $V$. Then there is a unique vector $v \in V$ such that*

$$\varphi(u) = \langle u, v \rangle$$

*for every $u \in V$.*

PROOF:    First we show that there exists a vector $v \in V$ such that $\varphi(u) = \langle u, v \rangle$ for every $u \in V$. Let $(e_1, \dots, e_n)$ be an orthonormal basis of $V$. Then

$$\begin{aligned}
\varphi(u) &= \varphi(\langle u, e_1 \rangle e_1 + \cdots + \langle u, e_n \rangle e_n) \\
&= \langle u, e_1 \rangle \varphi(e_1) + \cdots + \langle u, e_n \rangle \varphi(e_n) \\
&= \langle u, \overline{\varphi(e_1)} e_1 + \cdots + \overline{\varphi(e_n)} e_n \rangle
\end{aligned}$$

for every $u \in V$, where the first equality comes from 6.17. Thus setting $v = \overline{\varphi(e_1)} e_1 + \cdots + \overline{\varphi(e_n)} e_n$, we have $\varphi(u) = \langle u, v \rangle$ for every $u \in V$, as desired.

Now we prove that only one vector $v \in V$ has the desired behavior. Suppose $v_1, v_2 \in V$ are such that

$$\varphi(u) = \langle u, v_1 \rangle = \langle u, v_2 \rangle$$

for every $u \in V$. Then

$$0 = \langle u, v_1 \rangle - \langle u, v_2 \rangle = \langle u, v_1 - v_2 \rangle$$

for every $u \in V$. Taking $u = v_1 - v_2$ shows that $v_1 - v_2 = 0$. In other words, $v_1 = v_2$, completing the proof of the uniqueness part of the theorem. ∎

In addition to $V$, we need another finite-dimensional inner-product space.

> Let's agree that for the rest of this chapter
> $W$ is a finite-dimensional, nonzero, inner-product space over **F**.

*The word **adjoint** has another meaning in linear algebra. We will not need the second meaning, related to inverses, in this book. Just in case you encountered the second meaning for adjoint elsewhere, be warned that the two meanings for adjoint are unrelated to one another.*

Let $T \in \mathcal{L}(V, W)$. The **adjoint** of $T$, denoted $T^*$, is the function from $W$ to $V$ defined as follows. Fix $w \in W$. Consider the linear functional on $V$ that maps $v \in V$ to $\langle Tv, w \rangle$. Let $T^*w$ be the unique vector in $V$ such that this linear functional is given by taking inner products with $T^*w$ (6.45 guarantees the existence and uniqueness of a vector in $V$ with this property). In other words, $T^*w$ is the unique vector in $V$ such that

$$\langle Tv, w \rangle = \langle v, T^*w \rangle$$

for all $v \in V$.

Let's work out an example of how the adjoint is computed. Define $T : \mathbf{R}^3 \to \mathbf{R}^2$ by

$$T(x_1, x_2, x_3) = (x_2 + 3x_3, 2x_1).$$

Thus $T^*$ will be a function from $\mathbf{R}^2$ to $\mathbf{R}^3$. To compute $T^*$, fix a point $(y_1, y_2) \in \mathbf{R}^2$. Then

$$\begin{aligned}\langle (x_1, x_2, x_3), T^*(y_1, y_2) \rangle &= \langle T(x_1, x_2, x_3), (y_1, y_2) \rangle \\ &= \langle (x_2 + 3x_3, 2x_1), (y_1, y_2) \rangle \\ &= x_2 y_1 + 3x_3 y_1 + 2x_1 y_2 \\ &= \langle (x_1, x_2, x_3), (2y_2, y_1, 3y_1) \rangle\end{aligned}$$

for all $(x_1, x_2, x_3) \in \mathbf{R}^3$. This shows that

$$T^*(y_1, y_2) = (2y_2, y_1, 3y_1).$$

Note that in the example above, $T^*$ turned out to be not just a function from $\mathbf{R}^2$ to $\mathbf{R}^3$, but a linear map. That is true in general. Specifically, if $T \in \mathcal{L}(V, W)$, then $T^* \in \mathcal{L}(W, V)$. To prove this, suppose $T \in \mathcal{L}(V, W)$. Let's begin by checking additivity. Fix $w_1, w_2 \in W$. Then

*Adjoints play a crucial role in the important results in the next chapter.*

$$\begin{aligned}
\langle Tv, w_1 + w_2 \rangle &= \langle Tv, w_1 \rangle + \langle Tv, w_2 \rangle \\
&= \langle v, T^*w_1 \rangle + \langle v, T^*w_2 \rangle \\
&= \langle v, T^*w_1 + T^*w_2 \rangle,
\end{aligned}$$

which shows that $T^*w_1 + T^*w_2$ plays the role required of $T^*(w_1 + w_2)$. Because only one vector can behave that way, we must have

$$T^*w_1 + T^*w_2 = T^*(w_1 + w_2).$$

Now let's check the homogeneity of $T^*$. If $a \in \mathbf{F}$, then

$$\begin{aligned}
\langle Tv, aw \rangle &= \bar{a} \langle Tv, w \rangle \\
&= \bar{a} \langle v, T^*w \rangle \\
&= \langle v, aT^*w \rangle,
\end{aligned}$$

which shows that $aT^*w$ plays the role required of $T^*(aw)$. Because only one vector can behave that way, we must have

$$aT^*w = T^*(aw).$$

Thus $T^*$ is a linear map, as claimed.

You should verify that the function $T \mapsto T^*$ has the following properties:

**additivity**
   $(S + T)^* = S^* + T^*$ for all $S, T \in \mathcal{L}(V, W)$;

**conjugate homogeneity**
   $(aT)^* = \bar{a}T^*$ for all $a \in \mathbf{F}$ and $T \in \mathcal{L}(V, W)$;

**adjoint of adjoint**
   $(T^*)^* = T$ for all $T \in \mathcal{L}(V, W)$;

**identity**
   $I^* = I$, where $I$ is the identity operator on $V$;

**products**

$(ST)^* = T^*S^*$ for all $T \in \mathcal{L}(V, W)$ and $S \in \mathcal{L}(W, U)$ (here $U$ is an inner-product space over **F**).

The next result shows the relationship between the null space and the range of a linear map and its adjoint. The symbol $\Longleftrightarrow$ means "if and only if"; this symbol could also be read to mean "is equivalent to".

**6.46   Proposition:**  *Suppose $T \in \mathcal{L}(V, W)$. Then*

(a)      null $T^* = (\text{range } T)^{\perp}$;

(b)      range $T^* = (\text{null } T)^{\perp}$;

(c)      null $T = (\text{range } T^*)^{\perp}$;

(d)      range $T = (\text{null } T^*)^{\perp}$.

PROOF:   Let's begin by proving (a). Let $w \in W$. Then

$$
\begin{aligned}
w \in \text{null } T^* &\Longleftrightarrow T^*w = 0 \\
&\Longleftrightarrow \langle v, T^*w \rangle = 0 \text{ for all } v \in V \\
&\Longleftrightarrow \langle Tv, w \rangle = 0 \text{ for all } v \in V \\
&\Longleftrightarrow w \in (\text{range } T)^{\perp}.
\end{aligned}
$$

Thus null $T^* = (\text{range } T)^{\perp}$, proving (a).

If we take the orthogonal complement of both sides of (a), we get (d), where we have used 6.33. Finally, replacing $T$ with $T^*$ in (a) and (d) gives (c) and (b).                                              ∎

*If $\mathbf{F} = \mathbf{R}$, then the conjugate transpose of a matrix is the same as its **transpose**, which is the matrix obtained by interchanging the rows and columns.*

The ***conjugate transpose*** of an $m$-by-$n$ matrix is the $n$-by-$m$ matrix obtained by interchanging the rows and columns and then taking the complex conjugate of each entry. For example, the conjugate transpose of

$$
\begin{bmatrix} 2 & 3+4i & 7 \\ 6 & 5 & 8i \end{bmatrix}
$$

is the matrix

$$
\begin{bmatrix} 2 & 6 \\ 3-4i & 5 \\ 7 & -8i \end{bmatrix}.
$$

The next proposition shows how to compute the matrix of $T^*$ from the matrix of $T$. Caution: the proposition below applies only when we are dealing with orthonormal bases—with respect to nonorthonormal bases, the matrix of $T^*$ does not necessarily equal the conjugate transpose of the matrix of $T$.

**6.47** **Proposition:** *Suppose $T \in \mathcal{L}(V, W)$. If $(e_1, \ldots, e_n)$ is an orthonormal basis of $V$ and $(f_1, \ldots, f_m)$ is an orthonormal basis of $W$, then*

$$\mathcal{M}(T^*, (f_1, \ldots, f_m), (e_1, \ldots, e_n))$$

*is the conjugate transpose of*

$$\mathcal{M}(T, (e_1, \ldots, e_n), (f_1, \ldots, f_m)).$$

*The adjoint of a linear map does not depend on a choice of basis. This explains why we will emphasize adjoints of linear maps instead of conjugate transposes of matrices.*

PROOF: Suppose that $(e_1, \ldots, e_n)$ is an orthonormal basis of $V$ and $(f_1, \ldots, f_m)$ is an orthonormal basis of $W$. We write $\mathcal{M}(T)$ instead of the longer expression $\mathcal{M}(T, (e_1, \ldots, e_n), (f_1, \ldots, f_m))$; we also write $\mathcal{M}(T^*)$ instead of $\mathcal{M}(T^*, (f_1, \ldots, f_m), (e_1, \ldots, e_n))$.

Recall that we obtain the $k^{\text{th}}$ column of $\mathcal{M}(T)$ by writing $Te_k$ as a linear combination of the $f_j$'s; the scalars used in this linear combination then become the $k^{\text{th}}$ column of $\mathcal{M}(T)$. Because $(f_1, \ldots, f_m)$ is an orthonormal basis of $W$, we know how to write $Te_k$ as a linear combination of the $f_j$'s (see 6.17):

$$Te_k = \langle Te_k, f_1 \rangle f_1 + \cdots + \langle Te_k, f_m \rangle f_m.$$

Thus the entry in row $j$, column $k$, of $\mathcal{M}(T)$ is $\langle Te_k, f_j \rangle$. Replacing $T$ with $T^*$ and interchanging the roles played by the $e$'s and $f$'s, we see that the entry in row $j$, column $k$, of $\mathcal{M}(T^*)$ is $\langle T^* f_k, e_j \rangle$, which equals $\langle f_k, Te_j \rangle$, which equals $\overline{\langle Te_j, f_k \rangle}$, which equals the complex conjugate of the entry in row $k$, column $j$, of $\mathcal{M}(T)$. In other words, $\mathcal{M}(T^*)$ equals the conjugate transpose of $\mathcal{M}(T)$. ∎

## *Exercises*

1.    Prove that if $x, y$ are nonzero vectors in $\mathbf{R}^2$, then

$$\langle x, y \rangle = \|x\|\|y\| \cos \theta,$$

where $\theta$ is the angle between $x$ and $y$ (thinking of $x$ and $y$ as arrows with initial point at the origin). *Hint:* draw the triangle formed by $x$, $y$, and $x - y$; then use the law of cosines.

2.    Suppose $u, v \in V$. Prove that $\langle u, v \rangle = 0$ if and only if

$$\|u\| \leq \|u + av\|$$

for all $a \in \mathbf{F}$.

3.    Prove that

$$\left( \sum_{j=1}^{n} a_j b_j \right)^2 \leq \left( \sum_{j=1}^{n} j a_j{}^2 \right) \left( \sum_{j=1}^{n} \frac{b_j{}^2}{j} \right)$$

for all real numbers $a_1, \ldots, a_n$ and $b_1, \ldots, b_n$.

4.    Suppose $u, v \in V$ are such that

$$\|u\| = 3, \quad \|u + v\| = 4, \quad \|u - v\| = 6.$$

What number must $\|v\|$ equal?

5.    Prove or disprove: there is an inner product on $\mathbf{R}^2$ such that the associated norm is given by

$$\|(x_1, x_2)\| = |x_1| + |x_2|$$

for all $(x_1, x_2) \in \mathbf{R}^2$.

6.    Prove that if $V$ is a real inner-product space, then

$$\langle u, v \rangle = \frac{\|u + v\|^2 - \|u - v\|^2}{4}$$

for all $u, v \in V$.

7.    Prove that if $V$ is a complex inner-product space, then

$$\langle u, v \rangle = \frac{\|u + v\|^2 - \|u - v\|^2 + \|u + iv\|^2 i - \|u - iv\|^2 i}{4}$$

for all $u, v \in V$.

8.  A norm on a vector space $U$ is a function $\| \ \| : U \to [0, \infty)$ such that $\|u\| = 0$ if and only if $u = 0$, $\|\alpha u\| = |\alpha| \|u\|$ for all $\alpha \in \mathbf{F}$ and all $u \in U$, and $\|u + v\| \le \|u\| + \|v\|$ for all $u, v \in U$. Prove that a norm satisfying the parallelogram equality comes from an inner product (in other words, show that if $\| \ \|$ is a norm on $U$ satisfying the parallelogram equality, then there is an inner product $\langle \ , \ \rangle$ on $U$ such that $\|u\| = \langle u, u \rangle^{1/2}$ for all $u \in U$).

9.  Suppose $n$ is a positive integer. Prove that

$$\left( \frac{1}{\sqrt{2\pi}}, \frac{\sin x}{\sqrt{\pi}}, \frac{\sin 2x}{\sqrt{\pi}}, \dots, \frac{\sin nx}{\sqrt{\pi}}, \frac{\cos x}{\sqrt{\pi}}, \frac{\cos 2x}{\sqrt{\pi}}, \dots, \frac{\cos nx}{\sqrt{\pi}} \right)$$

    is an orthonormal list of vectors in $C[-\pi, \pi]$, the vector space of continuous real-valued functions on $[-\pi, \pi]$ with inner product

$$\langle f, g \rangle = \int_{-\pi}^{\pi} f(x)g(x)\, dx.$$

    *This orthonormal list is often used for modeling periodic phenomena such as tides.*

10. On $\mathcal{P}_2(\mathbf{R})$, consider the inner product given by

$$\langle p, q \rangle = \int_0^1 p(x)q(x)\, dx.$$

    Apply the Gram-Schmidt procedure to the basis $(1, x, x^2)$ to produce an orthonormal basis of $\mathcal{P}_2(\mathbf{R})$.

11. What happens if the Gram-Schmidt procedure is applied to a list of vectors that is not linearly independent?

12. Suppose $V$ is a real inner-product space and $(v_1, \dots, v_m)$ is a linearly independent list of vectors in $V$. Prove that there exist exactly $2^m$ orthonormal lists $(e_1, \dots, e_m)$ of vectors in $V$ such that

$$\text{span}(v_1, \dots, v_j) = \text{span}(e_1, \dots, e_j)$$

    for all $j \in \{1, \dots, m\}$.

13. Suppose $(e_1, \dots, e_m)$ is an orthonormal list of vectors in $V$. Let $v \in V$. Prove that

$$\|v\|^2 = |\langle v, e_1 \rangle|^2 + \dots + |\langle v, e_m \rangle|^2$$

    if and only if $v \in \text{span}(e_1, \dots, e_m)$.

14. Find an orthonormal basis of $P_2(\mathbf{R})$ (with inner product as in Exercise 10) such that the differentiation operator (the operator that takes $p$ to $p'$) on $P_2(\mathbf{R})$ has an upper-triangular matrix with respect to this basis.

15. Suppose $U$ is a subspace of $V$. Prove that
$$\dim U^\perp = \dim V - \dim U.$$

16. Suppose $U$ is a subspace of $V$. Prove that $U^\perp = \{0\}$ if and only if $U = V$.

17. Prove that if $P \in \mathcal{L}(V)$ is such that $P^2 = P$ and every vector in null $P$ is orthogonal to every vector in range $P$, then $P$ is an orthogonal projection.

18. Prove that if $P \in \mathcal{L}(V)$ is such that $P^2 = P$ and
$$\|Pv\| \le \|v\|$$
for every $v \in V$, then $P$ is an orthogonal projection.

19. Suppose $T \in \mathcal{L}(V)$ and $U$ is a subspace of $V$. Prove that $U$ is invariant under $T$ if and only if $P_U T P_U = T P_U$.

20. Suppose $T \in \mathcal{L}(V)$ and $U$ is a subspace of $V$. Prove that $U$ and $U^\perp$ are both invariant under $T$ if and only if $P_U T = T P_U$.

21. In $\mathbf{R}^4$, let
$$U = \mathrm{span}((1,1,0,0),(1,1,1,2)).$$
Find $u \in U$ such that $\|u - (1,2,3,4)\|$ is as small as possible.

22. Find $p \in P_3(\mathbf{R})$ such that $p(0) = 0$, $p'(0) = 0$, and
$$\int_0^1 |2 + 3x - p(x)|^2\,dx$$
is as small as possible.

23. Find $p \in P_5(\mathbf{R})$ that makes
$$\int_{-\pi}^{\pi} |\sin x - p(x)|^2\,dx$$
as small as possible. (The polynomial 6.40 is an excellent approximation to the answer to this exercise, but here you are asked to find the exact solution, which involves powers of $\pi$. A computer that can perform symbolic integration will be useful.)

24. Find a polynomial $q \in \mathcal{P}_2(\mathbf{R})$ such that

$$p(\frac{1}{2}) = \int_0^1 p(x)q(x)\,dx$$

for every $p \in \mathcal{P}_2(\mathbf{R})$.

25. Find a polynomial $q \in \mathcal{P}_2(\mathbf{R})$ such that

$$\int_0^1 p(x)(\cos \pi x)\,dx = \int_0^1 p(x)q(x)\,dx$$

for every $p \in \mathcal{P}_2(\mathbf{R})$.

26. Fix a vector $v \in V$ and define $T \in \mathcal{L}(V, \mathbf{F})$ by $Tu = \langle u, v \rangle$. For $a \in \mathbf{F}$, find a formula for $T^*a$.

27. Suppose $n$ is a positive integer. Define $T \in \mathcal{L}(\mathbf{F}^n)$ by

$$T(z_1, \ldots, z_n) = (0, z_1, \ldots, z_{n-1}).$$

Find a formula for $T^*(z_1, \ldots, z_n)$.

28. Suppose $T \in \mathcal{L}(V)$ and $\lambda \in \mathbf{F}$. Prove that $\lambda$ is an eigenvalue of $T$ if and only if $\bar{\lambda}$ is an eigenvalue of $T^*$.

29. Suppose $T \in \mathcal{L}(V)$ and $U$ is a subspace of $V$. Prove that $U$ is invariant under $T$ if and only if $U^\perp$ is invariant under $T^*$.

30. Suppose $T \in \mathcal{L}(V, W)$. Prove that

    (a)    $T$ is injective if and only if $T^*$ is surjective;

    (b)    $T$ is surjective if and only if $T^*$ is injective.

31. Prove that

$$\dim \operatorname{null} T^* = \dim \operatorname{null} T + \dim W - \dim V$$

and

$$\dim \operatorname{range} T^* = \dim \operatorname{range} T$$

for every $T \in \mathcal{L}(V, W)$.

32. Suppose $A$ is an $m$-by-$n$ matrix of real numbers. Prove that the dimension of the span of the columns of $A$ (in $\mathbf{R}^m$) equals the dimension of the span of the rows of $A$ (in $\mathbf{R}^n$).

# CHAPTER 7

# *Operators on Inner-Product Spaces*

The deepest results related to inner-product spaces deal with the subject to which we now turn—operators on inner-product spaces. By exploiting properties of the adjoint, we will develop a detailed description of several important classes of operators on inner-product spaces.

> Recall that **F** denotes **R** or **C**.
> Let's agree that for this chapter
> *V* is a finite-dimensional, nonzero, inner-product space over **F**.

127

# Self-Adjoint and Normal Operators

*Instead of self-adjoint,*
*some mathematicians*
*use the term **Hermitian***
*(in honor of the French*
*mathematician Charles*
*Hermite, who in 1873*
*published the first*
*proof that e is not the*
*root of any polynomial*
*with integer*
*coefficients).*

An operator $T \in \mathcal{L}(V)$ is called **self-adjoint** if $T = T^*$. For example, if $T$ is the operator on $\mathbf{F}^2$ whose matrix (with respect to the standard basis) is

$$\begin{bmatrix} 2 & b \\ 3 & 7 \end{bmatrix},$$

then $T$ is self-adjoint if and only if $b = 3$ (because $\mathcal{M}(T) = \mathcal{M}(T^*)$ if and only if $b = 3$; recall that $\mathcal{M}(T^*)$ is the conjugate transpose of $\mathcal{M}(T)$—see 6.47).

You should verify that the sum of two self-adjoint operators is self-adjoint and that the product of a real scalar and a self-adjoint operator is self-adjoint.

A good analogy to keep in mind (especially when $\mathbf{F} = \mathbf{C}$) is that the adjoint on $\mathcal{L}(V)$ plays a role similar to complex conjugation on $\mathbf{C}$. A complex number $z$ is real if and only if $z = \bar{z}$; thus a self-adjoint operator $(T = T^*)$ is analogous to a real number. We will see that this analogy is reflected in some important properties of self-adjoint operators, beginning with eigenvalues.

*If $\mathbf{F} = \mathbf{R}$, then by*
*definition every*
*eigenvalue is real, so*
*this proposition is*
*interesting only when*
*$\mathbf{F} = \mathbf{C}$.*

**7.1   Proposition:**   *Every eigenvalue of a self-adjoint operator is real.*

PROOF:   Suppose $T$ is a self-adjoint operator on $V$. Let $\lambda$ be an eigenvalue of $T$, and let $v$ be a nonzero vector in $V$ such that $Tv = \lambda v$. Then

$$\begin{aligned}
\lambda \|v\|^2 &= \langle \lambda v, v \rangle \\
&= \langle Tv, v \rangle \\
&= \langle v, Tv \rangle \\
&= \langle v, \lambda v \rangle \\
&= \bar{\lambda} \|v\|^2.
\end{aligned}$$

Thus $\lambda = \bar{\lambda}$, which means that $\lambda$ is real, as desired.   ∎

The next proposition is false for real inner-product spaces. As an example, consider the operator $T \in \mathcal{L}(\mathbf{R}^2)$ that is a counterclockwise rotation of 90° around the origin; thus $T(x, y) = (-y, x)$. Obviously $Tv$ is orthogonal to $v$ for every $v \in \mathbf{R}^2$, even though $T$ is not 0.

**7.2    Proposition:** *If $V$ is a complex inner-product space and $T$ is an operator on $V$ such that*

$$\langle Tv, v \rangle = 0$$

*for all $v \in V$, then $T = 0$.*

PROOF: Suppose $V$ is a complex inner-product space and $T \in \mathcal{L}(V)$. Then

$$\langle Tu, w \rangle = \frac{\langle T(u+w), u+w \rangle - \langle T(u-w), u-w \rangle}{4}$$
$$+ \frac{\langle T(u+iw), u+iw \rangle - \langle T(u-iw), u-iw \rangle}{4}i$$

for all $u, w \in V$, as can be verified by computing the right side. Note that each term on the right side is of the form $\langle Tv, v \rangle$ for appropriate $v \in V$. If $\langle Tv, v \rangle = 0$ for all $v \in V$, then the equation above implies that $\langle Tu, w \rangle = 0$ for all $u, w \in V$. This implies that $T = 0$ (take $w = Tu$). ∎

The following corollary is false for real inner-product spaces, as shown by considering any operator on a real inner-product space that is not self-adjoint.

**7.3    Corollary:**  *Let $V$ be a complex inner-product space and let $T \in \mathcal{L}(V)$. Then $T$ is self-adjoint if and only if*

$$\langle Tv, v \rangle \in \mathbf{R}$$

*for every $v \in V$.*

*This corollary provides another example of how self-adjoint operators behave like real numbers.*

PROOF: Let $v \in V$. Then

$$\begin{aligned}
\langle Tv, v \rangle - \overline{\langle Tv, v \rangle} &= \langle Tv, v \rangle - \langle v, Tv \rangle \\
&= \langle Tv, v \rangle - \langle T^*v, v \rangle \\
&= \langle (T - T^*)v, v \rangle.
\end{aligned}$$

If $\langle Tv, v \rangle \in \mathbf{R}$ for every $v \in V$, then the left side of the equation above equals 0, so $\langle (T - T^*)v, v \rangle = 0$ for every $v \in V$. This implies that $T - T^* = 0$ (by 7.2), and hence $T$ is self-adjoint.

Conversely, if $T$ is self-adjoint, then the right side of the equation above equals 0, so $\langle Tv, v \rangle = \overline{\langle Tv, v \rangle}$ for every $v \in V$. This implies that $\langle Tv, v \rangle \in \mathbf{R}$ for every $v \in V$, as desired. ∎

On a real inner-product space $V$, a nonzero operator $T$ may satisfy $\langle Tv, v \rangle = 0$ for all $v \in V$. However, the next proposition shows that this cannot happen for a self-adjoint operator.

**7.4    Proposition:** *If $T$ is a self-adjoint operator on $V$ such that*

$$\langle Tv, v \rangle = 0$$

*for all $v \in V$, then $T = 0$.*

PROOF:  We have already proved this (without the hypothesis that $T$ is self-adjoint) when $V$ is a complex inner-product space (see 7.2). Thus we can assume that $V$ is a real inner-product space and that $T$ is a self-adjoint operator on $V$. For $u, w \in V$, we have

**7.5**      $\langle Tu, w \rangle = \dfrac{\langle T(u + w), u + w \rangle - \langle T(u - w), u - w \rangle}{4};$

this is proved by computing the right side, using

$$\langle Tw, u \rangle = \langle w, Tu \rangle$$
$$= \langle Tu, w \rangle,$$

where the first equality holds because $T$ is self-adjoint and the second equality holds because we are working on a real inner-product space. If $\langle Tv, v \rangle = 0$ for all $v \in V$, then 7.5 implies that $\langle Tu, w \rangle = 0$ for all $u, w \in V$. This implies that $T = 0$ (take $w = Tu$).    ∎

An operator on an inner-product space is called ***normal*** if it commutes with its adjoint; in other words, $T \in \mathcal{L}(V)$ is normal if

$$TT^* = T^*T.$$

Obviously every self-adjoint operator is normal. For an example of a normal operator that is not self-adjoint, consider the operator on $\mathbf{F}^2$ whose matrix (with respect to the standard basis) is

$$\begin{bmatrix} 2 & -3 \\ 3 & 2 \end{bmatrix}.$$

Clearly this operator is not self-adjoint, but an easy calculation (which you should do) shows that it is normal.

We will soon see why normal operators are worthy of special attention. The next proposition provides a simple characterization of normal operators.

**7.6    Proposition:** *An operator $T \in \mathcal{L}(V)$ is normal if and only if*

$$\|Tv\| = \|T^*v\|$$

*for all $v \in V$.*

*Note that this proposition implies that* null $T =$ null $T^*$ *for every normal operator $T$.*

PROOF:   Let $T \in \mathcal{L}(V)$. We will prove both directions of this result at the same time. Note that

$$
\begin{aligned}
T \text{ is normal} &\iff T^*T - TT^* = 0 \\
&\iff \langle (T^*T - TT^*)v, v \rangle = 0 \quad \text{for all } v \in V \\
&\iff \langle T^*Tv, v \rangle = \langle TT^*v, v \rangle \quad \text{for all } v \in V \\
&\iff \|Tv\|^2 = \|T^*v\|^2 \quad \text{for all } v \in V,
\end{aligned}
$$

where we used 7.4 to establish the second equivalence (note that the operator $T^*T - TT^*$ is self-adjoint). The equivalence of the first and last conditions above gives the desired result.    ∎

Compare the next corollary to Exercise 28 in the previous chapter. That exercise implies that the eigenvalues of the adjoint of any operator are equal (as a set) to the complex conjugates of the eigenvalues of the operator. The exercise says nothing about eigenvectors because an operator and its adjoint may have different eigenvectors. However, the next corollary implies that a normal operator and its adjoint have the same eigenvectors.

**7.7    Corollary:** *Suppose $T \in \mathcal{L}(V)$ is normal. If $v \in V$ is an eigenvector of $T$ with eigenvalue $\lambda \in \mathbf{F}$, then $v$ is also an eigenvector of $T^*$ with eigenvalue $\bar{\lambda}$.*

PROOF:   Suppose $v \in V$ is an eigenvector of $T$ with eigenvalue $\lambda$. Thus $(T - \lambda I)v = 0$. Because $T$ is normal, so is $T - \lambda I$, as you should verify. Using 7.6, we have

$$0 = \|(T - \lambda I)v\| = \|(T - \lambda I)^*v\| = \|(T^* - \bar{\lambda}I)v\|,$$

and hence $v$ is an eigenvector of $T^*$ with eigenvalue $\bar{\lambda}$, as desired.    ∎

Because every self-adjoint operator is normal, the next result applies in particular to self-adjoint operators.

**7.8**  **Corollary:**  *If $T \in \mathcal{L}(V)$ is normal, then eigenvectors of $T$ corresponding to distinct eigenvalues are orthogonal.*

PROOF:  Suppose $T \in \mathcal{L}(V)$ is normal and $\alpha, \beta$ are distinct eigenvalues of $T$, with corresponding eigenvectors $u, v$. Thus $Tu = \alpha u$ and $Tv = \beta v$. From 7.7 we have $T^*v = \bar\beta v$. Thus

$$(\alpha - \beta)\langle u, v\rangle = \langle \alpha u, v\rangle - \langle u, \bar\beta v\rangle$$
$$= \langle Tu, v\rangle - \langle u, T^*v\rangle$$
$$= 0.$$

Because $\alpha \neq \beta$, the equation above implies that $\langle u, v\rangle = 0$. Thus $u$ and $v$ are orthogonal, as desired. ∎

## *The Spectral Theorem*

Recall that a diagonal matrix is a square matrix that is 0 everywhere except possibly along the diagonal. Recall also that an operator on $V$ has a diagonal matrix with respect to some basis if and only if there is a basis of $V$ consisting of eigenvectors of the operator (see 5.21).

The nicest operators on $V$ are those for which there is an *orthonormal* basis of $V$ with respect to which the operator has a diagonal matrix. These are precisely the operators $T \in \mathcal{L}(V)$ such that there is an orthonormal basis of $V$ consisting of eigenvectors of $T$. Our goal in this section is to prove the spectral theorem, which characterizes these operators as the normal operators when $\mathbf{F} = \mathbf{C}$ and as the self-adjoint operators when $\mathbf{F} = \mathbf{R}$. The spectral theorem is probably the most useful tool in the study of operators on inner-product spaces.

Because the conclusion of the spectral theorem depends on $\mathbf{F}$, we will break the spectral theorem into two pieces, called the complex spectral theorem and the real spectral theorem. As is often the case in linear algebra, complex vector spaces are easier to deal with than real vector spaces, so we present the complex spectral theorem first.

As an illustration of the complex spectral theorem, consider the normal operator $T \in \mathcal{L}(\mathbf{C}^2)$ whose matrix (with respect to the standard basis) is

$$\begin{bmatrix} 2 & -3 \\ 3 & 2 \end{bmatrix}.$$

You should verify that

$$\left( \frac{(i,1)}{\sqrt{2}}, \frac{(-i,1)}{\sqrt{2}} \right)$$

is an orthonormal basis of $\mathbf{C}^2$ consisting of eigenvectors of $T$ and that with respect to this basis, the matrix of $T$ is the diagonal matrix

$$\begin{bmatrix} 2+3i & 0 \\ 0 & 2-3i \end{bmatrix}.$$

**7.9   Complex Spectral Theorem:**   *Suppose that $V$ is a complex inner-product space and $T \in \mathcal{L}(V)$. Then $V$ has an orthonormal basis consisting of eigenvectors of $T$ if and only if $T$ is normal.*

PROOF: First suppose that $V$ has an orthonormal basis consisting of eigenvectors of $T$. With respect to this basis, $T$ has a diagonal matrix. The matrix of $T^*$ (with respect to the same basis) is obtained by taking the conjugate transpose of the matrix of $T$; hence $T^*$ also has a diagonal matrix. Any two diagonal matrices commute; thus $T$ commutes with $T^*$, which means that $T$ must be normal, as desired.

To prove the other direction, now suppose that $T$ is normal. There is an orthonormal basis $(e_1, \ldots, e_n)$ of $V$ with respect to which $T$ has an upper-triangular matrix (by 6.28). Thus we can write

*Because every self-adjoint operator is normal, the complex spectral theorem implies that every self-adjoint operator on a finite-dimensional complex inner-product space has a diagonal matrix with respect to some orthonormal basis.*

**7.10**      $$\mathcal{M}(T,(e_1,\ldots,e_n)) = \begin{bmatrix} a_{1,1} & \cdots & a_{1,n} \\ & \ddots & \vdots \\ 0 & & a_{n,n} \end{bmatrix}.$$

We will show that this matrix is actually a diagonal matrix, which means that $(e_1, \ldots, e_n)$ is an orthonormal basis of $V$ consisting of eigenvectors of $T$.

We see from the matrix above that

$$\|Te_1\|^2 = |a_{1,1}|^2$$

and

$$\|T^*e_1\|^2 = |a_{1,1}|^2 + |a_{1,2}|^2 + \cdots + |a_{1,n}|^2.$$

Because $T$ is normal, $\|Te_1\| = \|T^*e_1\|$ (see 7.6). Thus the two equations above imply that all entries in the first row of the matrix in 7.10, except possibly the first entry $a_{1,1}$, equal 0.

Now from 7.10 we see that

$$\|Te_2\|^2 = |a_{2,2}|^2$$

(because $a_{1,2} = 0$, as we showed in the paragraph above) and

$$\|T^*e_2\|^2 = |a_{2,2}|^2 + |a_{2,3}|^2 + \cdots + |a_{2,n}|^2.$$

Because $T$ is normal, $\|Te_2\| = \|T^*e_2\|$. Thus the two equations above imply that all entries in the second row of the matrix in 7.10, except possibly the diagonal entry $a_{2,2}$, equal 0.

Continuing in this fashion, we see that all the nondiagonal entries in the matrix 7.10 equal 0, as desired.                                      ∎

We will need two lemmas for our proof of the real spectral theorem. You could guess that the next lemma is true and even discover its proof by thinking about quadratic polynomials with real coefficients. Specifically, suppose $\alpha, \beta \in \mathbf{R}$ and $\alpha^2 < 4\beta$. Let $x$ be a real number. Then

*This technique of completing the square can be used to derive the quadratic formula.*

$$x^2 + \alpha x + \beta = (x + \frac{\alpha}{2})^2 + (\beta - \frac{\alpha^2}{4})$$
$$> 0.$$

In particular, $x^2 + \alpha x + \beta$ is an invertible real number (a convoluted way of saying that it is not 0). Replacing the real number $x$ with a self-adjoint operator (recall the analogy between real numbers and self-adjoint operators), we are led to the lemma below.

**7.11   Lemma:** *Suppose $T \in \mathcal{L}(V)$ is self-adjoint. If $\alpha, \beta \in \mathbf{R}$ are such that $\alpha^2 < 4\beta$, then*

$$T^2 + \alpha T + \beta I$$

*is invertible.*

PROOF:  Suppose $\alpha, \beta \in \mathbf{R}$ are such that $\alpha^2 < 4\beta$. Let $v$ be a nonzero vector in $V$. Then

$$\begin{aligned}
\langle (T^2 + \alpha T + \beta I)v, v \rangle &= \langle T^2 v, v \rangle + \alpha \langle Tv, v \rangle + \beta \langle v, v \rangle \\
&= \langle Tv, Tv \rangle + \alpha \langle Tv, v \rangle + \beta \|v\|^2 \\
&\geq \|Tv\|^2 - |\alpha| \|Tv\| \|v\| + \beta \|v\|^2 \\
&= (\|Tv\| - \frac{|\alpha| \|v\|}{2})^2 + (\beta - \frac{\alpha^2}{4}) \|v\|^2 \\
&> 0,
\end{aligned}$$

where the first inequality holds by the Cauchy-Schwarz inequality (6.6). The last inequality implies that $(T^2 + \alpha T + \beta I)v \neq 0$. Thus $T^2 + \alpha T + \beta I$ is injective, which implies that it is invertible (see 3.21).                  ∎

We have proved that every operator, self-adjoint or not, on a finite-dimensional complex vector space has an eigenvalue (see 5.10), so the next lemma tells us something new only for real inner-product spaces.

**7.12   Lemma:**   *Suppose $T \in \mathcal{L}(V)$ is self-adjoint. Then $T$ has an eigenvalue.*

PROOF:   As noted above, we can assume that $V$ is a real inner-product space. Let $n = \dim V$ and choose $v \in V$ with $v \neq 0$. Then

$$(v, Tv, T^2v, \ldots, T^nv)$$

cannot be linearly independent because $V$ has dimension $n$ and we have $n + 1$ vectors. Thus there exist real numbers $a_0, \ldots, a_n$, not all 0, such that

$$0 = a_0 v + a_1 Tv + \cdots + a_n T^n v.$$

*Here we are imitating the proof that $T$ has an invariant subspace of dimension 1 or 2 (see 5.24).*

Make the $a$'s the coefficients of a polynomial, which can be written in factored form (see 4.14) as

$$
\begin{aligned}
a_0 + a_1 x + &\cdots + a_n x^n \\
&= c(x^2 + \alpha_1 x + \beta_1) \ldots (x^2 + \alpha_M x + \beta_M)(x - \lambda_1) \ldots (x - \lambda_m),
\end{aligned}
$$

where $c$ is a nonzero real number, each $\alpha_j$, $\beta_j$, and $\lambda_j$ is real, each $\alpha_j{}^2 < 4\beta_j$, $m + M \geq 1$, and the equation holds for all real $x$. We then have

$$
\begin{aligned}
0 &= a_0 v + a_1 Tv + \cdots + a_n T^n v \\
&= (a_0 I + a_1 T + \cdots + a_n T^n)v \\
&= c(T^2 + \alpha_1 T + \beta_1 I) \ldots (T^2 + \alpha_M T + \beta_M I)(T - \lambda_1 I) \ldots (T - \lambda_m I)v.
\end{aligned}
$$

Each $T^2 + \alpha_j T + \beta_j I$ is invertible because $T$ is self-adjoint and each $\alpha_j{}^2 < 4\beta_j$ (see 7.11). Recall also that $c \neq 0$. Thus the equation above implies that

$$0 = (T - \lambda_1 I) \ldots (T - \lambda_m I)v.$$

Hence $T - \lambda_j I$ is not injective for at least one $j$. In other words, $T$ has an eigenvalue.                                                 ∎

As an illustration of the real spectral theorem, consider the self-adjoint operator $T$ on $\mathbf{R}^3$ whose matrix (with respect to the standard basis) is

$$
\begin{bmatrix}
14 & -13 & 8 \\
-13 & 14 & 8 \\
8 & 8 & -7
\end{bmatrix}.
$$

You should verify that

$$
\left( \frac{(1,-1,0)}{\sqrt{2}}, \frac{(1,1,1)}{\sqrt{3}}, \frac{(1,1,-2)}{\sqrt{6}} \right)
$$

is an orthonormal basis of $\mathbf{R}^3$ consisting of eigenvectors of $T$ and that with respect to this basis, the matrix of $T$ is the diagonal matrix

$$
\begin{bmatrix}
27 & 0 & 0 \\
0 & 9 & 0 \\
0 & 0 & -15
\end{bmatrix}.
$$

Combining the complex spectral theorem and the real spectral theorem, we conclude that every self-adjoint operator on $V$ has a diagonal matrix with respect to some orthonormal basis. This statement, which is the most useful part of the spectral theorem, holds regardless of whether $\mathbf{F} = \mathbf{C}$ or $\mathbf{F} = \mathbf{R}$.

**7.13   Real Spectral Theorem:**  *Suppose that $V$ is a real inner-product space and $T \in \mathcal{L}(V)$. Then $V$ has an orthonormal basis consisting of eigenvectors of $T$ if and only if $T$ is self-adjoint.*

PROOF:  First suppose that $V$ has an orthonormal basis consisting of eigenvectors of $T$. With respect to this basis, $T$ has a diagonal matrix. This matrix equals its conjugate transpose. Hence $T = T^*$ and so $T$ is self-adjoint, as desired.

To prove the other direction, now suppose that $T$ is self-adjoint. We will prove that $V$ has an orthonormal basis consisting of eigenvectors of $T$ by induction on the dimension of $V$. To get started, note that our desired result clearly holds if $\dim V = 1$. Now assume that $\dim V > 1$ and that the desired result holds on vector spaces of smaller dimension.

The idea of the proof is to take any eigenvector $u$ of $T$ with norm 1, then adjoin to it an orthonormal basis of eigenvectors of $T|_{\{u\}^\perp}$. Now for the details, the most important of which is verifying that $T|_{\{u\}^\perp}$ is self-adjoint (this allows us to apply our induction hypothesis).

Let $\lambda$ be any eigenvalue of $T$ (because $T$ is self-adjoint, we know from the previous lemma that it has an eigenvalue) and let $u \in V$ denote a corresponding eigenvector with $\|u\| = 1$. Let $U$ denote the one-dimensional subspace of $V$ consisting of all scalar multiples of $u$. Note that a vector $v \in V$ is in $U^\perp$ if and only if $\langle u, v \rangle = 0$.

*To get an eigenvector of norm* 1, *take any nonzero eigenvector and divide it by its norm.*

Suppose $v \in U^\perp$. Then because $T$ is self-adjoint, we have

$$\langle u, Tv \rangle = \langle Tu, v \rangle = \langle \lambda u, v \rangle = \lambda \langle u, v \rangle = 0,$$

and hence $Tv \in U^\perp$. Thus $Tv \in U^\perp$ whenever $v \in U^\perp$. In other words, $U^\perp$ is invariant under $T$. Thus we can define an operator $S \in \mathcal{L}(U^\perp)$ by $S = T|_{U^\perp}$. If $v, w \in U^\perp$, then

$$\langle Sv, w \rangle = \langle Tv, w \rangle = \langle v, Tw \rangle = \langle v, Sw \rangle,$$

which shows that $S$ is self-adjoint (note that in the middle equality above we used the self-adjointness of $T$). Thus, by our induction hypothesis, there is an orthonormal basis of $U^\perp$ consisting of eigenvectors of $S$. Clearly every eigenvector of $S$ is an eigenvector of $T$ (because $Sv = Tv$ for every $v \in U^\perp$). Thus adjoining $u$ to an orthonormal basis of $U^\perp$ consisting of eigenvectors of $S$ gives an orthonormal basis of $V$ consisting of eigenvectors of $T$, as desired. ∎

For $T \in \mathcal{L}(V)$ self-adjoint (or, more generally, $T \in \mathcal{L}(V)$ normal when $\mathbf{F} = \mathbf{C}$), the corollary below provides the nicest possible decomposition of $V$ into subspaces invariant under $T$. On each null$(T - \lambda_j I)$, the operator $T$ is just multiplication by $\lambda_j$.

**7.14   Corollary:** *Suppose that $T \in \mathcal{L}(V)$ is self-adjoint (or that $\mathbf{F} = \mathbf{C}$ and that $T \in \mathcal{L}(V)$ is normal). Let $\lambda_1, \ldots, \lambda_m$ denote the distinct eigenvalues of $T$. Then*

$$V = \text{null}(T - \lambda_1 I) \oplus \cdots \oplus \text{null}(T - \lambda_m I).$$

*Furthermore, each vector in each* null$(T - \lambda_j I)$ *is orthogonal to all vectors in the other subspaces of this decomposition.*

PROOF:   The spectral theorem (7.9 and 7.13) implies that $V$ has a basis consisting of eigenvectors of $T$. The desired decomposition of $V$ now follows from 5.21.

The orthogonality statement follows from 7.8. ∎

LIVERPOOL
JOHN MOORES UNIVERSITY
AVRIL ROBARTS LRC
TEL. 0151 231 4022

# Normal Operators on Real Inner-Product Spaces

The complex spectral theorem (7.9) gives a complete description of normal operators on complex inner-product spaces. In this section we will give a complete description of normal operators on real inner-product spaces. Along the way, we will encounter a proposition (7.18) and a technique (block diagonal matrices) that are useful for both real and complex inner-product spaces.

We begin with a description of the operators on a two-dimensional real inner-product space that are normal but not self-adjoint.

**7.15   Lemma:**  *Suppose V is a two-dimensional real inner-product space and $T \in \mathcal{L}(V)$. Then the following are equivalent:*

(a)    *T is normal but not self-adjoint;*

(b)    *the matrix of T with respect to every orthonormal basis of V has the form*

$$\begin{bmatrix} a & -b \\ b & a \end{bmatrix},$$

with $b \neq 0$;

(c)    *the matrix of T with respect to some orthonormal basis of V has the form*

$$\begin{bmatrix} a & -b \\ b & a \end{bmatrix},$$

with $b > 0$.

PROOF:   First suppose that (a) holds, so that $T$ is normal but not self-adjoint. Let $(e_1, e_2)$ be an orthonormal basis of $V$. Suppose

**7.16**               $$\mathcal{M}(T, (e_1, e_2)) = \begin{bmatrix} a & c \\ b & d \end{bmatrix}.$$

Then $\|Te_1\|^2 = a^2 + b^2$ and $\|T^*e_1\|^2 = a^2 + c^2$. Because $T$ is normal, $\|Te_1\| = \|T^*e_1\|$ (see 7.6); thus these equations imply that $b^2 = c^2$. Thus $c = b$ or $c = -b$. But $c \neq b$ because otherwise $T$ would be self-adjoint, as can be seen from the matrix in 7.16. Hence $c = -b$, so

**7.17**               $$\mathcal{M}(T, (e_1, e_2)) = \begin{bmatrix} a & -b \\ b & d \end{bmatrix}.$$

Of course, the matrix of $T^*$ is the transpose of the matrix above. Use matrix multiplication to compute the matrices of $TT^*$ and $T^*T$ (do it now). Because $T$ is normal, these two matrices must be equal. Equating the entries in the upper-right corner of the two matrices you computed, you will discover that $bd = ab$. Now $b \neq 0$ because otherwise $T$ would be self-adjoint, as can be seen from the matrix in 7.17. Thus $d = a$, completing the proof that (a) implies (b).

Now suppose that (b) holds. We want to prove that (c) holds. Choose any orthonormal basis $(e_1, e_2)$ of $V$. We know that the matrix of $T$ with respect to this basis has the form given by (b), with $b \neq 0$. If $b > 0$, then (c) holds and we have proved that (b) implies (c). If $b < 0$, then, as you should verify, the matrix of $T$ with respect to the orthonormal basis $(e_1, -e_2)$ equals $\begin{bmatrix} a & b \\ -b & a \end{bmatrix}$, where $-b > 0$; thus in this case we also see that (b) implies (c).

Now suppose that (c) holds, so that the matrix of $T$ with respect to some orthonormal basis has the form given in (c) with $b > 0$. Clearly the matrix of $T$ is not equal to its transpose (because $b \neq 0$), and hence $T$ is not self-adjoint. Now use matrix multiplication to verify that the matrices of $TT^*$ and $T^*T$ are equal. We conclude that $TT^* = T^*T$, and hence $T$ is normal. Thus (c) implies (a), completing the proof.  ∎

As an example of the notation we will use to write a matrix as a matrix of smaller matrices, consider the matrix

$$D = \begin{bmatrix} 1 & 1 & 2 & 2 & 2 \\ 1 & 1 & 2 & 2 & 2 \\ 0 & 0 & 3 & 3 & 3 \\ 0 & 0 & 3 & 3 & 3 \\ 0 & 0 & 3 & 3 & 3 \end{bmatrix}.$$

We can write this matrix in the form

$$D = \begin{bmatrix} A & B \\ 0 & C \end{bmatrix},$$

*Often we can understand a matrix better by thinking of it as composed of smaller matrices. We will use this technique in the next proposition and in later chapters.*

where

$$A = \begin{bmatrix} 1 & 1 \\ 1 & 1 \end{bmatrix}, \quad B = \begin{bmatrix} 2 & 2 & 2 \\ 2 & 2 & 2 \end{bmatrix}, \quad C = \begin{bmatrix} 3 & 3 & 3 \\ 3 & 3 & 3 \\ 3 & 3 & 3 \end{bmatrix},$$

and 0 denotes the 3-by-2 matrix consisting of all 0's.

The next result will play a key role in our characterization of the normal operators on a real inner-product space.

<table>
<tr><td>

*Without normality, an easier result also holds: if $T \in \mathcal{L}(V)$ and $U$ invariant under $T$, then $U^\perp$ is invariant under $T^*$; see Exercise 29 in Chapter 6.*

</td><td>

**7.18  Proposition:** *Suppose $T \in \mathcal{L}(V)$ is normal and $U$ is a subspace of $V$ that is invariant under $T$. Then*

(a)     $U^\perp$ *is invariant under $T$;*

(b)     $U$ *is invariant under $T^*$;*

(c)     $(T|_U)^* = (T^*)|_U$;

(d)     $T|_U$ *is a normal operator on $U$;*

(e)     $T|_{U^\perp}$ *is a normal operator on $U^\perp$.*

</td></tr>
</table>

PROOF:   First we will prove (a). Let $(e_1, \ldots, e_m)$ be an orthonormal basis of $U$. Extend to an orthonormal basis $(e_1, \ldots, e_m, f_1, \ldots, f_n)$ of $V$ (this is possible by 6.25). Because $U$ is invariant under $T$, each $Te_j$ is a linear combination of $(e_1, \ldots, e_m)$. Thus the matrix of $T$ with respect to the basis $(e_1, \ldots, e_m, f_1, \ldots, f_n)$ is of the form

$$\mathcal{M}(T) = \begin{array}{c} \\ \begin{array}{c} e_1 \\ \vdots \\ e_m \\ f_1 \\ \vdots \\ f_n \end{array} \end{array} \begin{array}{c} e_1 \ \cdots \ e_m \ f_1 \ \cdots \ f_n \\ \left[ \begin{array}{cc} A & B \\ & \\ 0 & C \end{array} \right] \end{array} ;$$

here $A$ denotes an $m$-by-$m$ matrix, 0 denotes the $n$-by-$m$ matrix consisting of all 0's, $B$ denotes an $m$-by-$n$ matrix, $C$ denotes an $n$-by-$n$ matrix, and for convenience the basis has been listed along the top and left sides of the matrix.

For each $j \in \{1, \ldots, m\}$, $\|Te_j\|^2$ equals the sum of the squares of the absolute values of the entries in the $j^{\text{th}}$ column of $A$ (see 6.17). Hence

**7.19**    $\displaystyle\sum_{j=1}^{m} \|Te_j\|^2 =$  the sum of the squares of the absolute values of the entries of $A$.

For each $j \in \{1, \ldots, m\}$, $\|T^*e_j\|^2$ equals the sum of the squares of the absolute values of the entries in the $j^{\text{th}}$ rows of $A$ and $B$. Hence

**7.20**    $\displaystyle\sum_{j=1}^{m} \|T^* e_j\|^2 = $    the sum of the squares of the absolute values of the entries of $A$ and $B$.

Because $T$ is normal, $\|Te_j\| = \|T^* e_j\|$ for each $j$ (see 7.6); thus

$$\sum_{j=1}^{m} \|Te_j\|^2 = \sum_{j=1}^{m} \|T^* e_j\|^2.$$

This equation, along with 7.19 and 7.20, implies that the sum of the squares of the absolute values of the entries of $B$ must equal 0. In other words, $B$ must be the matrix consisting of all 0's. Thus

**7.21**    $\mathcal{M}(T) = \begin{array}{c} \\ \\ \begin{array}{cc} e_1 & \cdots \ \ e_m \ f_1 \ \cdots \ f_n \end{array} \\ \begin{array}{c} e_1 \\ \vdots \\ e_m \\ f_1 \\ \vdots \\ f_n \end{array} \left[ \begin{array}{cc} A & 0 \\ \\ \\ 0 & C \end{array} \right] \end{array}.$

This representation shows that $Tf_k$ is in the span of $(f_1, \ldots, f_n)$ for each $k$. Because $(f_1, \ldots, f_n)$ is a basis of $U^\perp$, this implies that $Tv \in U^\perp$ whenever $v \in U^\perp$. In other words, $U^\perp$ is invariant under $T$, completing the proof of (a).

To prove (b), note that $\mathcal{M}(T^*)$ has a block of 0's in the lower left corner (because $\mathcal{M}(T)$, as given above, has a block of 0's in the upper right corner). In other words, each $T^* e_j$ can be written as a linear combination of $(e_1, \ldots, e_m)$. Thus $U$ is invariant under $T^*$, completing the proof of (b).

To prove (c), let $S = T|_U$. Fix $v \in U$. Then

$$\langle Su, v \rangle = \langle Tu, v \rangle$$
$$= \langle u, T^* v \rangle$$

for all $u \in U$. Because $T^* v \in U$ (by (b)), the equation above shows that $S^* v = T^* v$. In other words, $(T|_U)^* = (T^*)|_U$, completing the proof of (c).

To prove (d), note that $T$ commutes with $T^*$ (because $T$ is normal) and that $(T|_U)^* = (T^*)|_U$ (by (c)). Thus $T|_U$ commutes with its adjoint and hence is normal, completing the proof of (d).

To prove (e), note that in (d) we showed that the restriction of $T$ to any invariant subspace is normal. However, $U^\perp$ is invariant under $T$ (by (a)), and hence $T|_{U^\perp}$ is normal. ∎

In proving 7.18 we thought of a matrix as composed of smaller matrices. Now we need to make additional use of that idea. A **block diagonal matrix** is a square matrix of the form

*The key step in the proof of the last proposition was showing that $\mathcal{M}(T)$ is an appropriate block diagonal matrix; see 7.21.*

$$\begin{bmatrix} A_1 & & 0 \\ & \ddots & \\ 0 & & A_m \end{bmatrix},$$

where $A_1, \ldots, A_m$ are square matrices lying along the diagonal and all the other entries of the matrix equal 0. For example, the matrix

**7.22**
$$A = \begin{bmatrix} 4 & 0 & 0 & 0 & 0 \\ 0 & 2 & -3 & 0 & 0 \\ 0 & 3 & 2 & 0 & 0 \\ 0 & 0 & 0 & 1 & -7 \\ 0 & 0 & 0 & 7 & 1 \end{bmatrix}$$

is a block diagonal matrix with

$$A = \begin{bmatrix} A_1 & & 0 \\ & A_2 & \\ 0 & & A_3 \end{bmatrix},$$

where

**7.23**  $A_1 = \begin{bmatrix} 4 \end{bmatrix}, \quad A_2 = \begin{bmatrix} 2 & -3 \\ 3 & 2 \end{bmatrix}, \quad A_3 = \begin{bmatrix} 1 & -7 \\ 7 & 1 \end{bmatrix}.$

If $A$ and $B$ are block diagonal matrices of the form

$$A = \begin{bmatrix} A_1 & & 0 \\ & \ddots & \\ 0 & & A_m \end{bmatrix}, \quad B = \begin{bmatrix} B_1 & & 0 \\ & \ddots & \\ 0 & & B_m \end{bmatrix},$$

where $A_j$ has the same size as $B_j$ for $j = 1, \ldots, m$, then $AB$ is a block diagonal matrix of the form

**7.24**
$$AB = \begin{bmatrix} A_1 B_1 & & 0 \\ & \ddots & \\ 0 & & A_m B_m \end{bmatrix},$$

as you should verify. In other words, to multiply together two block diagonal matrices (with the same size blocks), just multiply together the corresponding entries on the diagonal, as with diagonal matrices.

A diagonal matrix is a special case of a block diagonal matrix where each block has size 1-by-1. At the other extreme, every square matrix is a block diagonal matrix because we can take the first (and only) block to be the entire matrix. Thus to say that an operator has a block diagonal matrix with respect to some basis tells us nothing unless we know something about the size of the blocks. The smaller the blocks, the nicer the operator (in the vague sense that the matrix then contains more 0's). The nicest situation is to have an orthonormal basis that gives a diagonal matrix. We have shown that this happens on a complex inner-product space precisely for the normal operators (see 7.9) and on a real inner-product space precisely for the self-adjoint operators (see 7.13).

*Note that if an operator T has a block diagonal matrix with respect to some basis, then the entry in any 1-by-1 block on the diagonal of this matrix must be an eigenvalue of T.*

Our next result states that each normal operator on a real inner-product space comes close to having a diagonal matrix—specifically, we get a block diagonal matrix with respect to some orthonormal basis, with each block having size at most 2-by-2. We cannot expect to do better than that because on a real inner-product space there exist normal operators that do not have a diagonal matrix with respect to any basis. For example, the operator $T \in \mathcal{L}(\mathbf{R}^2)$ defined by $T(x, y) = (-y, x)$ is normal (as you should verify) but has no eigenvalues; thus this particular $T$ does not have even an upper-triangular matrix with respect to any basis of $\mathbf{R}^2$.

Note that the matrix in 7.22 is the type of matrix promised by the theorem below. In particular, each block of 7.22 (see 7.23) has size at most 2-by-2 and each of the 2-by-2 blocks has the required form (upper left entry equals lower right entry, lower left entry is positive, and upper right entry equals the negative of lower left entry).

**7.25   Theorem:**   *Suppose that $V$ is a real inner-product space and $T \in \mathcal{L}(V)$. Then $T$ is normal if and only if there is an orthonormal basis of $V$ with respect to which $T$ has a block diagonal matrix where each block is a 1-by-1 matrix or a 2-by-2 matrix of the form*

**7.26**
$$\begin{bmatrix} a & -b \\ b & a \end{bmatrix},$$

*with $b > 0$.*

PROOF: To prove the easy direction, first suppose that there is an orthonormal basis of $V$ such that the matrix of $T$ is a block diagonal matrix where each block is a 1-by-1 matrix or a 2-by-2 matrix of the form 7.26. With respect to this basis, the matrix of $T$ commutes with the matrix of $T^*$ (which is the conjugate of the matrix of $T$), as you should verify (use formula 7.24 for the product of two block diagonal matrices). Thus $T$ commutes with $T^*$, which means that $T$ is normal.

To prove the other direction, now suppose that $T$ is normal. We will prove our desired result by induction on the dimension of $V$. To get started, note that our desired result clearly holds if $\dim V = 1$ (trivially) or if $\dim V = 2$ (if $T$ is self-adjoint, use the real spectral theorem 7.13; if $T$ is not self-adjoint, use 7.15).

Now assume that $\dim V > 2$ and that the desired result holds on vector spaces of smaller dimension. Let $U$ be a subspace of $V$ of dimension 1 that is invariant under $T$ if such a subspace exists (in other words, if $T$ has a nonzero eigenvector, let $U$ be the span of this eigenvector). If no such subspace exists, let $U$ be a subspace of $V$ of dimension 2 that is invariant under $T$ (an invariant subspace of dimension 1 or 2 always exists by 5.24).

*In a real vector space with dimension 1, there are precisely two vectors with norm 1.*

If $\dim U = 1$, choose a vector in $U$ with norm 1; this vector will be an orthonormal basis of $U$, and of course the matrix of $T|_U$ is a 1-by-1 matrix. If $\dim U = 2$, then $T|_U$ is normal (by 7.18) but not self-adjoint (otherwise $T|_U$, and hence $T$, would have a nonzero eigenvector; see 7.12), and thus we can choose an orthonormal basis of $U$ with respect to which the matrix of $T|_U$ has the form 7.26 (see 7.15).

Now $U^\perp$ is invariant under $T$ and $T|_{U^\perp}$ is a normal operator on $U^\perp$ (see 7.18). Thus by our induction hypothesis, there is an orthonormal basis of $U^\perp$ with respect to which the matrix of $T|_{U^\perp}$ has the desired form. Adjoining this basis to the basis of $U$ gives an orthonormal basis of $V$ with respect to which the matrix of $T$ has the desired form.  ∎

## *Positive Operators*

*Many mathematicians also use the term* **positive semidefinite operator**, *which means the same as positive operator.*

An operator $T \in \mathcal{L}(V)$ is called **positive** if $T$ is self-adjoint and

$$\langle Tv, v \rangle \geq 0$$

for all $v \in V$. Note that if $V$ is a complex vector space, then the condition that $T$ be self-adjoint can be dropped from this definition (by 7.3).

You should verify that every orthogonal projection is positive. For another set of examples, look at the proof of 7.11, where we showed that if $T \in \mathcal{L}(V)$ is self-adjoint and $\alpha, \beta \in \mathbf{R}$ are such that $\alpha^2 < 4\beta$, then $T^2 + \alpha T + \beta I$ is positive.

An operator $S$ is called a **square root** of an operator $T$ if $S^2 = T$. For example, if $T \in \mathcal{L}(\mathbf{F}^3)$ is defined by $T(z_1, z_2, z_3) = (z_3, 0, 0)$, then the operator $S \in \mathcal{L}(\mathbf{F}^3)$ defined by $S(z_1, z_2, z_3) = (z_2, z_3, 0)$ is a square root of $T$.

The following theorem is the main result about positive operators. Note that its characterizations of the positive operators correspond to characterizations of the nonnegative numbers among $\mathbf{C}$. Specifically, a complex number $z$ is nonnegative if and only if it has a nonnegative square root, corresponding to condition (c) below. Also, $z$ is nonnegative if and only if it has a real square root, corresponding to condition (d) below. Finally, $z$ is nonnegative if and only if there exists a complex number $w$ such that $z = \bar{w}w$, corresponding to condition (e) below.

*The positive operators correspond, in some sense, to the numbers $[0, \infty)$, so better terminology would call these nonnegative instead of positive. However, operator theorists consistently call these the positive operators, so we will follow that custom.*

**7.27** **Theorem:** *Let $T \in \mathcal{L}(V)$. Then the following are equivalent:*

(a)     *$T$ is positive;*

(b)     *$T$ is self-adjoint and all the eigenvalues of $T$ are nonnegative;*

(c)     *$T$ has a positive square root;*

(d)     *$T$ has a self-adjoint square root;*

(e)     *there exists an operator $S \in \mathcal{L}(V)$ such that $T = S^*S$.*

PROOF:   We will prove that (a) $\Rightarrow$ (b) $\Rightarrow$ (c) $\Rightarrow$ (d) $\Rightarrow$ (e) $\Rightarrow$ (a).

First suppose that (a) holds, so that $T$ is positive. Obviously $T$ is self-adjoint (by the definition of a positive operator). To prove the other condition in (b), suppose that $\lambda$ is an eigenvalue of $T$. Let $v$ be a nonzero eigenvector of $T$ corresponding to $\lambda$. Then

$$0 \leq \langle Tv, v \rangle$$
$$= \langle \lambda v, v \rangle$$
$$= \lambda \langle v, v \rangle,$$

and thus $\lambda$ is a nonnegative number. Hence (b) holds.

Now suppose that (b) holds, so that $T$ is self-adjoint and all the eigenvalues of $T$ are nonnegative. By the spectral theorem (7.9 and 7.13), there is an orthonormal basis $(e_1,\dots,e_n)$ of $V$ consisting of eigen-vectors of $T$. Let $\lambda_1,\dots,\lambda_n$ be the eigenvalues of $T$ corresponding to $e_1,\dots,e_n$, so that each $\lambda_j$ is a nonnegative number. Define $S \in \mathcal{L}(V)$ by

$$Se_j = \sqrt{\lambda_j}\,e_j$$

for $j = 1,\dots,n$. Then $S$ is a positive operator, as you should verify. Furthermore, $S^2 e_j = \lambda_j e_j = T e_j$ for each $j$, which implies that $S^2 = T$. Thus $S$ is a positive square root of $T$, and hence (c) holds.

Clearly (c) implies (d) (because, by definition, every positive operator is self-adjoint).

Now suppose that (d) holds, meaning that there exists a self-adjoint operator $S$ on $V$ such that $T = S^2$. Then $T = S^*S$ (because $S^* = S$), and hence (e) holds.

Finally, suppose that (e) holds. Let $S \in \mathcal{L}(V)$ be such that $T = S^*S$. Then $T^* = (S^*S)^* = S^*(S^*)^* = S^*S = T$, and hence $T$ is self-adjoint. To complete the proof that (a) holds, note that

$$\langle Tv, v \rangle = \langle S^*Sv, v \rangle$$
$$= \langle Sv, Sv \rangle$$
$$\geq 0$$

for every $v \in V$. Thus $T$ is positive. $\blacksquare$

Each nonnegative number has a unique nonnegative square root. The next proposition shows that positive operators enjoy a similar property. Because of this proposition, we can use the notation $\sqrt{T}$ to denote the unique positive square root of a positive operator $T$, just as $\sqrt{\lambda}$ denotes the unique nonnegative square root of a nonnegative number $\lambda$.

*A positive operator can have infinitely many square roots (though only one of them can be positive). For example, the identity operator on $V$ has infinitely many square roots if $\dim V > 1$.*

**7.28   Proposition:** *Every positive operator on $V$ has a unique positive square root.*

PROOF:  Suppose $T \in \mathcal{L}(V)$ is positive. Let $\lambda_1,\dots,\lambda_m$ denote the distinct eigenvalues of $T$; because $T$ is positive, all these numbers are nonnegative (by 7.27). Because $T$ is self-adjoint, we have

**7.29**          $V = \text{null}(T - \lambda_1 I) \oplus \cdots \oplus \text{null}(T - \lambda_m I);$

see 7.14.

Now suppose $S \in \mathcal{L}(V)$ is a positive square root of $T$. Suppose $\alpha$ is an eigenvalue of $S$. If $v \in \text{null}(S - \alpha I)$, then $Sv = \alpha v$, which implies that

**7.30**
$$Tv = S^2 v = \alpha^2 v,$$

so $v \in \text{null}(T - \alpha^2 I)$. Thus $\alpha^2$ is an eigenvalue of $T$, which means that $\alpha^2$ must equal some $\lambda_j$. In other words, $\alpha = \sqrt{\lambda_j}$ for some $j$. Furthermore, 7.30 implies that

**7.31**
$$\text{null}(S - \sqrt{\lambda_j} I) \subset \text{null}(T - \lambda_j I).$$

In the paragraph above, we showed that the only possible eigenvalues for $S$ are $\sqrt{\lambda_1}, \ldots, \sqrt{\lambda_m}$. Because $S$ is self-adjoint, this implies that

**7.32**
$$V = \text{null}(S - \sqrt{\lambda_1} I) \oplus \cdots \oplus \text{null}(S - \sqrt{\lambda_m} I);$$

see 7.14. Now 7.29, 7.32, and 7.31 imply that

$$\text{null}(S - \sqrt{\lambda_j} I) = \text{null}(T - \lambda_j I)$$

for each $j$. In other words, on $\text{null}(T - \lambda_j I)$, the operator $S$ is just multiplication by $\sqrt{\lambda_j}$. Thus $S$, the positive square root of $T$, is uniquely determined by $T$. ∎

## Isometries

An operator $S \in \mathcal{L}(V)$ is called an **isometry** if

$$\|Sv\| = \|v\|$$

for all $v \in V$. In other words, an operator is an isometry if it preserves norms. For example, $\lambda I$ is an isometry whenever $\lambda \in \mathbf{F}$ satisfies $|\lambda| = 1$. More generally, suppose $\lambda_1, \ldots, \lambda_n$ are scalars with absolute value 1 and $S \in \mathcal{L}(V)$ satisfies $S(e_j) = \lambda_j e_j$ for some orthonormal basis $(e_1, \ldots, e_n)$ of $V$. Suppose $v \in V$. Then

**7.33**
$$v = \langle v, e_1 \rangle e_1 + \cdots + \langle v, e_n \rangle e_n$$

and

**7.34**
$$\|v\|^2 = |\langle v, e_1 \rangle|^2 + \cdots + |\langle v, e_n \rangle|^2,$$

*The Greek word **isos** means equal; the Greek word **metron** means measure. Thus **isometry** literally means equal measure.*

where we have used 6.17. Applying $S$ to both sides of 7.33 gives

$$Sv = \langle v, e_1 \rangle Se_1 + \cdots + \langle v, e_n \rangle Se_n$$
$$= \lambda_1 \langle v, e_1 \rangle e_1 + \cdots + \lambda_n \langle v, e_n \rangle e_n.$$

The last equation, along with the equation $|\lambda_j| = 1$, shows that

**7.35**            $$\|Sv\|^2 = |\langle v, e_1 \rangle|^2 + \cdots + |\langle v, e_n \rangle|^2.$$

Comparing 7.34 and 7.35 shows that $\|v\| = \|Sv\|$. In other words, $S$ is an isometry.

For another example, let $\theta \in \mathbf{R}$. Then the operator on $\mathbf{R}^2$ of counterclockwise rotation (centered at the origin) by an angle of $\theta$ is an isometry (you should find the matrix of this operator with respect to the standard basis of $\mathbf{R}^2$).

If $S \in \mathcal{L}(V)$ is an isometry, then $S$ is injective (because if $Sv = 0$, then $\|v\| = \|Sv\| = 0$, and hence $v = 0$). Thus every isometry is invertible (by 3.21).

The next theorem provides several conditions that are equivalent to being an isometry. These equivalences have several important interpretations. In particular, the equivalence of (a) and (b) shows that an isometry preserves inner products. Because (a) implies (d), we see that if $S$ is an isometry and $(e_1, \ldots, e_n)$ is an orthonormal basis of $V$, then the columns of the matrix of $S$ (with respect to this basis) are orthonormal; because (e) implies (a), we see that the converse also holds. Because (a) is equivalent to conditions (i) and (j), we see that in the last sentence we can replace "columns" with "rows".

*An isometry on a real inner-product space is often called an **orthogonal** operator. An isometry on a complex inner-product space is often called a **unitary** operator. We will use the term isometry so that our results can apply to both real and complex inner-product spaces.*

**7.36   Theorem:** *Suppose $S \in \mathcal{L}(V)$. Then the following are equivalent:*

(a)   *$S$ is an isometry;*

(b)   *$\langle Su, Sv \rangle = \langle u, v \rangle$ for all $u, v \in V$;*

(c)   *$S^*S = I$;*

(d)   *$(Se_1, \ldots, Se_n)$ is orthonormal whenever $(e_1, \ldots, e_n)$ is an orthonormal list of vectors in $V$;*

(e)   *there exists an orthonormal basis $(e_1, \ldots, e_n)$ of $V$ such that $(Se_1, \ldots, Se_n)$ is orthonormal;*

(f)   *$S^*$ is an isometry;*

(g)     $\langle S^*u, S^*v \rangle = \langle u, v \rangle$ for all $u, v \in V$;

(h)     $SS^* = I$;

(i)     $(S^*e_1, \ldots, S^*e_n)$ is orthonormal whenever $(e_1, \ldots, e_n)$ is an orthonormal list of vectors in $V$;

(j)     there exists an orthonormal basis $(e_1, \ldots, e_n)$ of $V$ such that $(S^*e_1, \ldots, S^*e_n)$ is orthonormal.

PROOF:   First suppose that (a) holds. If $V$ is a real inner-product space, then for every $u, v \in V$ we have

$$\begin{aligned}
\langle Su, Sv \rangle &= (\|Su + Sv\|^2 - \|Su - Sv\|^2)/4 \\
&= (\|S(u + v)\|^2 - \|S(u - v)\|^2)/4 \\
&= (\|u + v\|^2 - \|u - v\|^2)/4 \\
&= \langle u, v \rangle,
\end{aligned}$$

where the first equality comes from Exercise 6 in Chapter 6, the second equality comes from the linearity of $S$, the third equality holds because $S$ is an isometry, and the last equality again comes from Exercise 6 in Chapter 6. If $V$ is a complex inner-product space, then use Exercise 7 in Chapter 6 instead of Exercise 6 to obtain the same conclusion. In either case, we see that (a) implies (b).

Now suppose that (b) holds. Then

$$\begin{aligned}
\langle (S^*S - I)u, v \rangle &= \langle Su, Sv \rangle - \langle u, v \rangle \\
&= 0
\end{aligned}$$

for every $u, v \in V$. Taking $v = (S^*S - I)u$, we see that $S^*S - I = 0$. Hence $S^*S = I$, proving that (b) implies (c).

Now suppose that (c) holds. Suppose $(e_1, \ldots, e_n)$ is an orthonormal list of vectors in $V$. Then

$$\begin{aligned}
\langle Se_j, Se_k \rangle &= \langle S^*Se_j, e_k \rangle \\
&= \langle e_j, e_k \rangle.
\end{aligned}$$

Hence $(Se_1, \ldots, Se_n)$ is orthonormal, proving that (c) implies (d).

Obviously (d) implies (e).

Now suppose (e) holds. Let $(e_1, \ldots, e_n)$ be an orthonormal basis of $V$ such that $(Se_1, \ldots, Se_n)$ is orthonormal. If $v \in V$, then

$$\|Sv\|^2 = \|S(\langle v, e_1 \rangle e_1 + \cdots + \langle v, e_n \rangle e_n)\|^2$$
$$= \|\langle v, e_1 \rangle Se_1 + \cdots + \langle v, e_n \rangle Se_n\|^2$$
$$= |\langle v, e_1 \rangle|^2 + \cdots + |\langle v, e_n \rangle|^2$$
$$= \|v\|^2,$$

where the first and last equalities come from 6.17. Taking square roots, we see that $S$ is an isometry, proving that (e) implies (a).

Having shown that (a) $\Rightarrow$ (b) $\Rightarrow$ (c) $\Rightarrow$ (d) $\Rightarrow$ (e) $\Rightarrow$ (a), we know at this stage that (a) through (e) are all equivalent to each other. Replacing $S$ with $S^*$, we see that (f) through (j) are all equivalent to each other. Thus to complete the proof, we need only show that one of the conditions in the group (a) through (e) is equivalent to one of the conditions in the group (f) through (j). The easiest way to connect the two groups of conditions is to show that (c) is equivalent to (h). In general, of course, $S$ need not commute with $S^*$. However, $S^*S = I$ if and only if $SS^* = I$; this is a special case of Exercise 23 in Chapter 3. Thus (c) is equivalent to (h), completing the proof.  ∎

The last theorem shows that every isometry is normal (see (a), (c), and (h) of 7.36). Thus the characterizations of normal operators can be used to give complete descriptions of isometries. We do this in the next two theorems.

**7.37  Theorem:**  *Suppose V is a complex inner-product space and $S \in \mathcal{L}(V)$. Then S is an isometry if and only if there is an orthonormal basis of V consisting of eigenvectors of S all of whose corresponding eigenvalues have absolute value 1.*

PROOF:  We already proved (see the first paragraph of this section) that if there is an orthonormal basis of $V$ consisting of eigenvectors of $S$ all of whose eigenvalues have absolute value 1, then $S$ is an isometry.

To prove the other direction, suppose $S$ is an isometry. By the complex spectral theorem (7.9), there is an orthonormal basis $(e_1, \ldots, e_n)$ of $V$ consisting of eigenvectors of $S$. For $j \in \{1, \ldots, n\}$, let $\lambda_j$ be the eigenvalue corresponding to $e_j$. Then

$$|\lambda_j| = \|\lambda_j e_j\| = \|Se_j\| = \|e_j\| = 1.$$

Thus each eigenvalue of $S$ has absolute value 1, completing the proof.  ∎

If $\theta \in \mathbf{R}$, then the operator on $\mathbf{R}^2$ of counterclockwise rotation (centered at the origin) by an angle of $\theta$ has matrix 7.39 with respect to the standard basis, as you should verify. The next result states that every isometry on a real inner-product space is composed of pieces that look like rotations on two-dimensional subspaces, pieces that equal the identity operator, and pieces that equal multiplication by $-1$.

**7.38   Theorem:**   *Suppose that $V$ is a real inner-product space and $S \in \mathcal{L}(V)$. Then $S$ is an isometry if and only if there is an orthonormal basis of $V$ with respect to which $S$ has a block diagonal matrix where each block on the diagonal is a 1-by-1 matrix containing 1 or $-1$ or a 2-by-2 matrix of the form*

**7.39**
$$\begin{bmatrix} \cos \theta & -\sin \theta \\ \sin \theta & \cos \theta \end{bmatrix},$$

*with $\theta \in (0, \pi)$.*

> *This theorem implies that an isometry on an odd-dimensional real inner-product space must have 1 or $-1$ as an eigenvalue.*

PROOF:   First suppose that $S$ is an isometry. Because $S$ is normal, there is an orthonormal basis of $V$ such that with respect to this basis $S$ has a block diagonal matrix, where each block is a 1-by-1 matrix or a 2-by-2 matrix of the form

**7.40**
$$\begin{bmatrix} a & -b \\ b & a \end{bmatrix},$$

with $b > 0$ (see 7.25).

If $\lambda$ is an entry in a 1-by-1 along the diagonal of the matrix of $S$ (with respect to the basis mentioned above), then there is a basis vector $e_j$ such that $Se_j = \lambda e_j$. Because $S$ is an isometry, this implies that $|\lambda| = 1$. Thus $\lambda = 1$ or $\lambda = -1$ because these are the only real numbers with absolute value 1.

Now consider a 2-by-2 matrix of the form 7.40 along the diagonal of the matrix of $S$. There are basis vectors $e_j, e_{j+1}$ such that

$$Se_j = ae_j + be_{j+1}.$$

Thus

$$1 = \|e_j\|^2 = \|Se_j\|^2 = a^2 + b^2.$$

The equation above, along with the condition $b > 0$, implies that there exists a number $\theta \in (0, \pi)$ such that $a = \cos \theta$ and $b = \sin \theta$. Thus the matrix 7.40 has the required form 7.39, completing the proof in this direction.

Conversely, now suppose that there is an orthonormal basis of $V$ with respect to which the matrix of $S$ has the form required by the theorem. Thus there is a direct sum decomposition

$$V = U_1 \oplus \cdots \oplus U_m,$$

where each $U_j$ is a subspace of $V$ of dimension 1 or 2. Furthermore, any two vectors belonging to distinct $U$'s are orthogonal, and each $S|_{U_j}$ is an isometry mapping $U_j$ into $U_j$. If $v \in V$, we can write

$$v = u_1 + \cdots + u_m,$$

where each $u_j \in U_j$. Applying $S$ to the equation above and then taking norms gives

$$\begin{aligned}
\|Sv\|^2 &= \|Su_1 + \cdots + Su_m\|^2 \\
&= \|Su_1\|^2 + \cdots + \|Su_m\|^2 \\
&= \|u_1\|^2 + \cdots + \|u_m\|^2 \\
&= \|v\|^2.
\end{aligned}$$

Thus $S$ is an isometry, as desired.                                    ∎

## *Polar and Singular-Value Decompositions*

Recall our analogy between $\mathbf{C}$ and $\mathcal{L}(V)$. Under this analogy, a complex number $z$ corresponds to an operator $T$, and $\bar{z}$ corresponds to $T^*$. The real numbers correspond to the self-adjoint operators, and the nonnegative numbers correspond to the (badly named) positive operators. Another distinguished subset of $\mathbf{C}$ is the unit circle, which consists of the complex numbers $z$ such that $|z| = 1$. The condition $|z| = 1$ is equivalent to the condition $\bar{z}z = 1$. Under our analogy, this would correspond to the condition $T^*T = I$, which is equivalent to $T$ being an isometry (see 7.36). In other words, the unit circle in $\mathbf{C}$ corresponds to the isometries.

Continuing with our analogy, note that each complex number $z$ except 0 can be written in the form

$$z = \left(\frac{z}{|z|}\right)|z| = \left(\frac{z}{|z|}\right)\sqrt{\bar{z}z},$$

where the first factor, namely, $z/|z|$, is an element of the unit circle. Our analogy leads us to guess that any operator $T \in \mathcal{L}(V)$ can be written as an isometry times $\sqrt{T^*T}$. That guess is indeed correct, as we now prove.

**7.41   Polar Decomposition:** *If $T \in \mathcal{L}(V)$, then there exists an isometry $S \in \mathcal{L}(V)$ such that*

$$T = S\sqrt{T^*T}.$$

PROOF:  Suppose $T \in \mathcal{L}(V)$. If $v \in V$, then

$$
\begin{aligned}
\|Tv\|^2 &= \langle Tv, Tv \rangle \\
&= \langle T^*Tv, v \rangle \\
&= \langle \sqrt{T^*T}\sqrt{T^*T}v, v \rangle \\
&= \langle \sqrt{T^*T}v, \sqrt{T^*T}v \rangle \\
&= \|\sqrt{T^*T}v\|^2.
\end{aligned}
$$

Thus

**7.42**
$$\|Tv\| = \|\sqrt{T^*T}v\|$$

for all $v \in V$.

Define a linear map $S_1$: range $\sqrt{T^*T} \to$ range $T$ by

**7.43**
$$S_1(\sqrt{T^*T}v) = Tv.$$

The idea of the proof is to extend $S_1$ to an isometry $S \in \mathcal{L}(V)$ such that $T = S\sqrt{T^*T}$. Now for the details.

First we must check that $S_1$ is well defined. To do this, suppose $v_1, v_2 \in V$ are such that $\sqrt{T^*T}v_1 = \sqrt{T^*T}v_2$. For the definition given by 7.43 to make sense, we must show that $Tv_1 = Tv_2$. However,

$$
\begin{aligned}
\|Tv_1 - Tv_2\| &= \|T(v_1 - v_2)\| \\
&= \|\sqrt{T^*T}(v_1 - v_2)\| \\
&= \|\sqrt{T^*T}v_1 - \sqrt{T^*T}v_2\| \\
&= 0,
\end{aligned}
$$

where the second equality holds by 7.42. The equation above shows that $Tv_1 = Tv_2$, so $S_1$ is indeed well defined. You should verify that $S_1$ is a linear map.

*If you know a bit of complex analysis, you will recognize the analogy to polar coordinates for complex numbers: every complex number can be written in the form $e^{\theta i}r$, where $\theta \in [0, 2\pi)$ and $r \geq 0$. Note that $e^{\theta i}$ is in the unit circle, corresponding to $S$ being an isometry, and $r$ is nonnegative, corresponding to $\sqrt{T^*T}$ being a positive operator.*

*In the rest of the proof all we are doing is extending $S_1$ to an isometry $S$ on all of $V$.*

We see from 7.43 that $S_1$ maps range $\sqrt{T^*T}$ onto range $T$. Clearly 7.42 and 7.43 imply that $\|S_1 u\| = \|u\|$ for all $u \in$ range $\sqrt{T^*T}$. In particular, $S_1$ is injective. Thus from 3.4, applied to $S_1$, we have

$$\dim \text{range}\, \sqrt{T^*T} = \dim \text{range}\, T.$$

This implies that $\dim(\text{range}\, \sqrt{T^*T})^{\perp} = \dim(\text{range}\, T)^{\perp}$ (see Exercise 15 in Chapter 6). Thus orthonormal bases $(e_1, \ldots, e_m)$ of $(\text{range}\, \sqrt{T^*T})^{\perp}$ and $(f_1, \ldots, f_m)$ of $(\text{range}\, T)^{\perp}$ can be chosen; the key point here is that these two orthonormal bases have the same length. Define a linear map $S_2 \colon (\text{range}\, \sqrt{T^*T})^{\perp} \to (\text{range}\, T)^{\perp}$ by

$$S_2(a_1 e_1 + \cdots + a_m e_m) = a_1 f_1 + \cdots + a_m f_m.$$

Obviously $\|S_2 w\| = \|w\|$ for all $w \in (\text{range}\, \sqrt{T^*T})^{\perp}$.

Now let $S$ be the operator on $V$ that equals $S_1$ on range $\sqrt{T^*T}$ and equals $S_2$ on $(\text{range}\, \sqrt{T^*T})^{\perp}$. More precisely, recall that each $v \in V$ can be written uniquely in the form

**7.44** $$v = u + w,$$

where $u \in$ range $\sqrt{T^*T}$ and $w \in (\text{range}\, \sqrt{T^*T})^{\perp}$ (see 6.29). For $v \in V$ with decomposition as above, define $Sv$ by

$$Sv = S_1 u + S_2 w.$$

For each $v \in V$ we have

$$S(\sqrt{T^*T}\, v) = S_1(\sqrt{T^*T}\, v) = Tv,$$

so $T = S\sqrt{T^*T}$, as desired. All that remains is to show that $S$ is an isometry. However, this follows easily from the two uses of the Pythagorean theorem: if $v \in V$ has decomposition as in 7.44, then

$$\begin{aligned}
\|Sv\|^2 &= \|S_1 u + S_2 w\|^2 \\
&= \|S_1 u\|^2 + \|S_2 w\|^2 \\
&= \|u\|^2 + \|w\|^2 \\
&= \|v\|^2,
\end{aligned}$$

where the second equality above holds because $S_1 u \in$ range $T$ and $S_2 u \in (\text{range}\, T)^{\perp}$. ∎

The polar decomposition (7.41) states that each operator on $V$ is the product of an isometry and a positive operator. Thus we can write each operator on $V$ as the product of two operators, each of which comes from a class that we have completely described and that we understand reasonably well. The isometries are described by 7.37 and 7.38; the positive operators (which are all self-adjoint) are described by the spectral theorem (7.9 and 7.13).

Specifically, suppose $T = S\sqrt{T^*T}$ is the polar decomposition of $T \in \mathcal{L}(V)$, where $S$ is an isometry. Then there is an orthonormal basis of $V$ with respect to which $S$ has a diagonal matrix (if $\mathbf{F} = \mathbf{C}$) or a block diagonal matrix with blocks of size at most 2-by-2 (if $\mathbf{F} = \mathbf{R}$), and there is an orthonormal basis of $V$ with respect to which $\sqrt{T^*T}$ has a diagonal matrix. Warning: there may not exist an orthonormal basis that simultaneously puts the matrices of both $S$ and $\sqrt{T^*T}$ into these nice forms (diagonal or block diagonal with small blocks). In other words, $S$ may require one orthonormal basis and $\sqrt{T^*T}$ may require a different orthonormal basis.

Suppose $T \in \mathcal{L}(V)$. The **singular values** of $T$ are the eigenvalues of $\sqrt{T^*T}$, with each eigenvalue $\lambda$ repeated $\dim \operatorname{null}(\sqrt{T^*T} - \lambda I)$ times. The singular values of $T$ are all nonnegative because they are the eigenvalues of the positive operator $\sqrt{T^*T}$.

For example, if $T \in \mathcal{L}(\mathbf{F}^4)$ is defined by

**7.45**            $T(z_1, z_2, z_3, z_4) = (0, 3z_1, 2z_2, -3z_4),$

then $T^*T(z_1, z_2, z_3, z_4) = (9z_1, 4z_2, 0, 9z_4)$, as you should verify. Thus

$$\sqrt{T^*T}(z_1, z_2, z_3, z_4) = (3z_1, 2z_2, 0, 3z_4),$$

and we see that the eigenvalues of $\sqrt{T^*T}$ are $3, 2, 0$. Clearly

$$\dim \operatorname{null}(\sqrt{T^*T} - 3I) = 2, \ \dim \operatorname{null}(\sqrt{T^*T} - 2I) = 1, \ \dim \operatorname{null}\sqrt{T^*T} = 1.$$

Hence the singular values of $T$ are $3, 3, 2, 0$. In this example $-3$ and $0$ are the only eigenvalues of $T$, as you should verify.

Each $T \in \mathcal{L}(V)$ has $\dim V$ singular values, as can be seen by applying the spectral theorem and 5.21 (see especially part (e)) to the positive (hence self-adjoint) operator $\sqrt{T^*T}$. For example, the operator $T$ defined by 7.45 on the four-dimensional vector space $\mathbf{F}^4$ has four singular values (they are $3, 3, 2, 0$), as we saw in the previous paragraph.

The next result shows that every operator on $V$ has a nice description in terms of its singular values and two orthonormal bases of $V$.

**7.46    Singular-Value Decomposition:**    *Suppose $T \in \mathcal{L}(V)$ has singular values $s_1, \ldots, s_n$. Then there exist orthonormal bases $(e_1, \ldots, e_n)$ and $(f_1, \ldots, f_n)$ of $V$ such that*

**7.47**                    $$Tv = s_1 \langle v, e_1 \rangle f_1 + \cdots + s_n \langle v, e_n \rangle f_n$$

*for every $v \in V$.*

PROOF:    By the spectral theorem (also see 7.14) applied to $\sqrt{T^*T}$, there is an orthonormal basis $(e_1, \ldots, e_n)$ of $V$ such that $\sqrt{T^*T} e_j = s_j e_j$ for $j = 1, \ldots, n$. We have

$$v = \langle v, e_1 \rangle e_1 + \cdots + \langle v, e_n \rangle e_n$$

for every $v \in V$ (see 6.17). Apply $\sqrt{T^*T}$ to both sides of this equation, getting

$$\sqrt{T^*T} v = s_1 \langle v, e_1 \rangle e_1 + \cdots + s_n \langle v, e_n \rangle e_n$$

*This proof illustrates the usefulness of the polar decomposition.*    for every $v \in V$. By the polar decomposition (see 7.41), there is an isometry $S \in \mathcal{L}(V)$ such that $T = S\sqrt{T^*T}$. Apply $S$ to both sides of the equation above, getting

$$Tv = s_1 \langle v, e_1 \rangle Se_1 + \cdots + s_n \langle v, e_n \rangle Se_n$$

for every $v \in V$. For each $j$, let $f_j = Se_j$. Because $S$ is an isometry, $(f_1, \ldots, f_n)$ is an orthonormal basis of $V$ (see 7.36). The equation above now becomes

$$Tv = s_1 \langle v, e_1 \rangle f_1 + \cdots + s_n \langle v, e_n \rangle f_n$$

for every $v \in V$, completing the proof.                    ∎

When we worked with linear maps from one vector space to a second vector space, we considered the matrix of a linear map with respect to a basis for the first vector space and a basis for the second vector space. When dealing with operators, which are linear maps from a vector space to itself, we almost always use only one basis, making it play both roles.

The singular-value decomposition allows us a rare opportunity to use two different bases for the matrix of an operator. To do this, suppose $T \in \mathcal{L}(V)$. Let $s_1, \ldots, s_n$ denote the singular values of $T$, and let $(e_1, \ldots, e_n)$ and $(f_1, \ldots, f_n)$ be orthonormal bases of $V$ such that the singular-value decomposition 7.47 holds. Then clearly

$$\mathcal{M}(T,(e_1,\ldots,e_n),(f_1,\ldots,f_n)) = \begin{bmatrix} s_1 & & 0 \\ & \ddots & \\ 0 & & s_n \end{bmatrix}.$$

In other words, every operator on $V$ has a diagonal matrix with respect to some orthonormal bases of $V$, provided that we are permitted to use two different bases rather than a single basis as customary when working with operators.

Singular values and the singular-value decomposition have many applications (some are given in the exercises), including applications in computational linear algebra. To compute numeric approximations to the singular values of an operator $T$, first compute $T^*T$ and then compute approximations to the eigenvalues of $T^*T$ (good techniques exist for approximating eigenvalues of positive operators). The nonnegative square roots of these (approximate) eigenvalues of $T^*T$ will be the (approximate) singular values of $T$ (as can be seen from the proof of 7.28). In other words, the singular values of $T$ can be approximated without computing the square root of $T^*T$.

## Exercises

1.    Make $P_2(\mathbf{R})$ into an inner-product space by defining

$$\langle p,q \rangle = \int_0^1 p(x)q(x)\,dx.$$

Define $T \in \mathcal{L}(P_2(\mathbf{R}))$ by $T(a_0 + a_1x + a_2x^2) = a_1x$.

(a)    Show that $T$ is not self-adjoint.

(b)    The matrix of $T$ with respect to the basis $(1, x, x^2)$ is

$$\begin{bmatrix} 0 & 0 & 0 \\ 0 & 1 & 0 \\ 0 & 0 & 0 \end{bmatrix}.$$

This matrix equals its conjugate transpose, even though $T$ is not self-adjoint. Explain why this is not a contradiction.

2.    Prove or give a counterexample: the product of any two self-adjoint operators on a finite-dimensional inner-product space is self-adjoint.

3.    (a)    Show that if $V$ is a real inner-product space, then the set of self-adjoint operators on $V$ is a subspace of $\mathcal{L}(V)$.

(b)    Show that if $V$ is a complex inner-product space, then the set of self-adjoint operators on $V$ is not a subspace of $\mathcal{L}(V)$.

4.    Suppose $P \in \mathcal{L}(V)$ is such that $P^2 = P$. Prove that $P$ is an orthogonal projection if and only if $P$ is self-adjoint.

5.    Show that if $\dim V \geq 2$, then the set of normal operators on $V$ is not a subspace of $\mathcal{L}(V)$.

6.    Prove that if $T \in \mathcal{L}(V)$ is normal, then

$$\text{range } T = \text{range } T^*.$$

7.    Prove that if $T \in \mathcal{L}(V)$ is normal, then

$$\text{null } T^k = \text{null } T \quad \text{and} \quad \text{range } T^k = \text{range } T$$

for every positive integer $k$.

8.  Prove that there does not exist a self-adjoint operator $T \in \mathcal{L}(\mathbf{R}^3)$ such that $T(1,2,3) = (0,0,0)$ and $T(2,5,7) = (2,5,7)$.

9.  Prove that a normal operator on a complex inner-product space is self-adjoint if and only if all its eigenvalues are real.

    *Exercise 9 strengthens the analogy (for normal operators) between self-adjoint operators and real numbers.*

10. Suppose $V$ is a complex inner-product space and $T \in \mathcal{L}(V)$ is a normal operator such that $T^9 = T^8$. Prove that $T$ is self-adjoint and $T^2 = T$.

11. Suppose $V$ is a complex inner-product space. Prove that every normal operator on $V$ has a square root. (An operator $S \in \mathcal{L}(V)$ is called a **square root** of $T \in \mathcal{L}(V)$ if $S^2 = T$.)

12. Give an example of a real inner-product space $V$ and $T \in \mathcal{L}(V)$ and real numbers $\alpha, \beta$ with $\alpha^2 < 4\beta$ such that $T^2 + \alpha T + \beta I$ is not invertible.

    *This exercise shows that the hypothesis that $T$ is self-adjoint is needed in 7.11, even for real vector spaces.*

13. Prove or give a counterexample: every self-adjoint operator on $V$ has a cube root. (An operator $S \in \mathcal{L}(V)$ is called a **cube root** of $T \in \mathcal{L}(V)$ if $S^3 = T$.)

14. Suppose $T \in \mathcal{L}(V)$ is self-adjoint, $\lambda \in \mathbf{F}$, and $\epsilon > 0$. Prove that if there exists $v \in V$ such that $\|v\| = 1$ and

    $$\|Tv - \lambda v\| < \epsilon,$$

    then $T$ has an eigenvalue $\lambda'$ such that $|\lambda - \lambda'| < \epsilon$.

15. Suppose $U$ is a finite-dimensional real vector space and $T \in \mathcal{L}(U)$. Prove that $U$ has a basis consisting of eigenvectors of $T$ if and only if there is an inner product on $U$ that makes $T$ into a self-adjoint operator.

16. Give an example of an operator $T$ on an inner product space such that $T$ has an invariant subspace whose orthogonal complement is not invariant under $T$.

    *This exercise shows that 7.18 can fail without the hypothesis that $T$ is normal.*

17. Prove that the sum of any two positive operators on $V$ is positive.

18. Prove that if $T \in \mathcal{L}(V)$ is positive, then so is $T^k$ for every positive integer $k$.

19. Suppose that $T$ is a positive operator on $V$. Prove that $T$ is invertible if and only if

$$\langle Tv, v \rangle > 0$$

for every $v \in V \setminus \{0\}$.

20. Prove or disprove: the identity operator on $\mathbf{F}^2$ has infinitely many self-adjoint square roots.

21. Prove or give a counterexample: if $S \in \mathcal{L}(V)$ and there exists an orthonormal basis $(e_1, \ldots, e_n)$ of $V$ such that $\|Se_j\| = 1$ for each $e_j$, then $S$ is an isometry.

22. Prove that if $S \in \mathcal{L}(\mathbf{R}^3)$ is an isometry, then there exists a nonzero vector $x \in \mathbf{R}^3$ such that $S^2 x = x$.

23. Define $T \in \mathcal{L}(\mathbf{F}^3)$ by

$$T(z_1, z_2, z_3) = (z_3, 2z_1, 3z_2).$$

Find (explicitly) an isometry $S \in \mathcal{L}(\mathbf{F}^3)$ such that $T = S\sqrt{T^*T}$.

*Exercise 24 shows that if we write $T$ as the product of an isometry and a positive operator (as in the polar decomposition), then the positive operator must equal $\sqrt{T^*T}$.*

24. Suppose $T \in \mathcal{L}(V)$, $S \in \mathcal{L}(V)$ is an isometry, and $R \in \mathcal{L}(V)$ is a positive operator such that $T = SR$. Prove that $R = \sqrt{T^*T}$.

25. Suppose $T \in \mathcal{L}(V)$. Prove that $T$ is invertible if and only if there exists a unique isometry $S \in \mathcal{L}(V)$ such that $T = S\sqrt{T^*T}$.

26. Prove that if $T \in \mathcal{L}(V)$ is self-adjoint, then the singular values of $T$ equal the absolute values of the eigenvalues of $T$ (repeated appropriately).

27. Prove or give a counterexample: if $T \in \mathcal{L}(V)$, then the singular values of $T^2$ equal the squares of the singular values of $T$.

28. Suppose $T \in \mathcal{L}(V)$. Prove that $T$ is invertible if and only if $0$ is not a singular value of $T$.

29. Suppose $T \in \mathcal{L}(V)$. Prove that $\dim \operatorname{range} T$ equals the number of nonzero singular values of $T$.

30. Suppose $S \in \mathcal{L}(V)$. Prove that $S$ is an isometry if and only if all the singular values of $S$ equal 1.

31.  Suppose $T_1, T_2 \in \mathcal{L}(V)$. Prove that $T_1$ and $T_2$ have the same singular values if and only if there exist isometries $S_1, S_2 \in \mathcal{L}(V)$ such that $T_1 = S_1 T_2 S_2$.

32.  Suppose $T \in \mathcal{L}(V)$ has singular-value decomposition given by

$$Tv = s_1 \langle v, e_1 \rangle f_1 + \cdots + s_n \langle v, e_n \rangle f_n$$

for every $v \in V$, where $s_1, \ldots, s_n$ are the singular values of $T$ and $(e_1, \ldots, e_n)$ and $(f_1, \ldots, f_n)$ are orthonormal bases of $V$.

(a)   Prove that

$$T^* v = s_1 \langle v, f_1 \rangle e_1 + \cdots + s_n \langle v, f_n \rangle e_n$$

for every $v \in V$.

(b)   Prove that if $T$ is invertible, then

$$T^{-1} v = \frac{\langle v, f_1 \rangle e_1}{s_1} + \cdots + \frac{\langle v, f_n \rangle e_n}{s_n}$$

for every $v \in V$.

33.  Suppose $T \in \mathcal{L}(V)$. Let $\hat{s}$ denote the smallest singular value of $T$, and let $s$ denote the largest singular value of $T$. Prove that

$$\hat{s} \|v\| \leq \|Tv\| \leq s \|v\|$$

for every $v \in V$.

34.  Suppose $T', T'' \in \mathcal{L}(V)$. Let $s'$ denote the largest singular value of $T'$, let $s''$ denote the largest singular value of $T''$, and let $s$ denote the largest singular value of $T' + T''$. Prove that $s \leq s' + s''$.

# CHAPTER 8

# *Operators on Complex Vector Spaces*

In this chapter we delve deeper into the structure of operators on complex vector spaces. An inner product does not help with this material, so we return to the general setting of a finite-dimensional vector space (as opposed to the more specialized context of an inner-product space). Thus our assumptions for this chapter are as follows:

Recall that $\mathbf{F}$ denotes $\mathbf{R}$ or $\mathbf{C}$.
Also, $V$ is a finite-dimensional, nonzero vector space over $\mathbf{F}$.

Some of the results in this chapter are valid on real vector spaces, so we have not assumed that $V$ is a complex vector space. Most of the results in this chapter that are proved only for complex vector spaces have analogous results on real vector spaces that are proved in the next chapter. We deal with complex vector spaces first because the proofs on complex vector spaces are often simpler than the analogous proofs on real vector spaces.

163

# *Generalized Eigenvectors*

Unfortunately some operators do not have enough eigenvectors to lead to a good description. Thus in this section we introduce the concept of generalized eigenvectors, which will play a major role in our description of the structure of an operator.

To understand why we need more than eigenvectors, let's examine the question of describing an operator by decomposing its domain into invariant subspaces. Fix $T \in \mathcal{L}(V)$. We seek to describe $T$ by finding a "nice" direct sum decomposition

**8.1**                           $$V = U_1 \oplus \cdots \oplus U_m,$$

where each $U_j$ is a subspace of $V$ invariant under $T$. The simplest possible nonzero invariant subspaces are one-dimensional. A decomposition 8.1 where each $U_j$ is a one-dimensional subspace of $V$ invariant under $T$ is possible if and only if $V$ has a basis consisting of eigenvectors of $T$ (see 5.21). This happens if and only if $V$ has the decomposition

**8.2**                 $$V = \text{null}(T - \lambda_1 I) \oplus \cdots \oplus \text{null}(T - \lambda_m I),$$

where $\lambda_1, \ldots, \lambda_m$ are the distinct eigenvalues of $T$ (see 5.21).

In the last chapter we showed that a decomposition of the form 8.2 holds for every self-adjoint operator on an inner-product space (see 7.14). Sadly, a decomposition of the form 8.2 may not hold for more general operators, even on a complex vector space. An example was given by the operator in 5.19, which does not have enough eigenvectors for 8.2 to hold. Generalized eigenvectors, which we now introduce, will remedy this situation. Our main goal in this chapter is to show that if $V$ is a complex vector space and $T \in \mathcal{L}(V)$, then

$$V = \text{null}(T - \lambda_1 I)^{\dim V} \oplus \cdots \oplus \text{null}(T - \lambda_m I)^{\dim V},$$

where $\lambda_1, \ldots, \lambda_m$ are the distinct eigenvalues of $T$ (see 8.23).

Suppose $T \in \mathcal{L}(V)$ and $\lambda$ is an eigenvalue of $T$. A vector $v \in V$ is called a ***generalized eigenvector*** of $T$ corresponding to $\lambda$ if

**8.3**                           $$(T - \lambda I)^j v = 0$$

for some positive integer $j$. Note that every eigenvector of $T$ is a generalized eigenvector of $T$ (take $j = 1$ in the equation above), but the converse is not true. For example, if $T \in \mathcal{L}(\mathbf{C}^3)$ is defined by

$$T(z_1, z_2, z_3) = (z_2, 0, z_3),$$

then $T^2(z_1, z_2, 0) = 0$ for all $z_1, z_2 \in \mathbf{C}$. Hence every element of $\mathbf{C}^3$ whose last coordinate equals 0 is a generalized eigenvector of $T$. As you should verify,

$$\mathbf{C}^3 = \{(z_1, z_2, 0) : z_1, z_2 \in \mathbf{C}\} \oplus \{(0, 0, z_3) : z_3 \in \mathbf{C}\},$$

where the first subspace on the right equals the set of generalized eigenvectors for this operator corresponding to the eigenvalue 0 and the second subspace on the right equals the set of generalized eigenvectors corresponding to the eigenvalue 1. Later in this chapter we will prove that a decomposition using generalized eigenvectors exists for every operator on a complex vector space (see 8.23).

Though $j$ is allowed to be an arbitrary integer in the definition of a generalized eigenvector, we will soon see that every generalized eigenvector satisfies an equation of the form 8.3 with $j$ equal to the dimension of $V$. To prove this, we now turn to a study of null spaces of powers of an operator.

*Note that we do not define the concept of a generalized eigenvalue because this would not lead to anything new. Reason: if $(T - \lambda I)^j$ is not injective for some positive integer $j$, then $T - \lambda I$ is not injective, and hence $\lambda$ is an eigenvalue of $T$.*

Suppose $T \in \mathcal{L}(V)$ and $k$ is a nonnegative integer. If $T^k v = 0$, then $T^{k+1}v = T(T^k v) = T(0) = 0$. Thus null $T^k \subset$ null $T^{k+1}$. In other words, we have

**8.4**     $\{0\} = \text{null } T^0 \subset \text{null } T^1 \subset \cdots \subset \text{null } T^k \subset \text{null } T^{k+1} \subset \cdots.$

The next proposition says that once two consecutive terms in this sequence of subspaces are equal, then all later terms in the sequence are equal.

**8.5**   **Proposition:**  *If $T \in \mathcal{L}(V)$ and $m$ is a nonnegative integer such that* null $T^m = $ null $T^{m+1}$, *then*

$$\text{null } T^0 \subset \text{null } T^1 \subset \cdots \subset \text{null } T^m = \text{null } T^{m+1} = \text{null } T^{m+2} = \cdots.$$

PROOF:   Suppose $T \in \mathcal{L}(V)$ and $m$ is a nonnegative integer such that null $T^m = $ null $T^{m+1}$. Let $k$ be a positive integer. We want to prove that

$$\text{null } T^{m+k} = \text{null } T^{m+k+1}.$$

We already know that null $T^{m+k} \subset$ null $T^{m+k+1}$. To prove the inclusion in the other direction, suppose that $v \in$ null $T^{m+k+1}$. Then

$$0 = T^{m+k+1}v = T^{m+1}(T^k v).$$

Hence

$$T^k v \in \text{null } T^{m+1} = \text{null } T^m.$$

Thus $0 = T^m(T^k v) = T^{m+k}v$, which means that $v \in \text{null } T^{m+k}$. This implies that null $T^{m+k+1} \subset \text{null } T^{m+k}$, completing the proof.  ∎

The proposition above raises the question of whether there must exist a nonnegative integer $m$ such that null $T^m = \text{null } T^{m+1}$. The proposition below shows that this equality holds at least when $m$ equals the dimension of the vector space on which $T$ operates.

**8.6    Proposition:**  *If $T \in \mathcal{L}(V)$, then*

$$\text{null } T^{\dim V} = \text{null } T^{\dim V+1} = \text{null } T^{\dim V+2} = \cdots.$$

PROOF:  Suppose $T \in \mathcal{L}(V)$. To get our desired conclusion, we need only prove that null $T^{\dim V} = \text{null } T^{\dim V+1}$ (by 8.5). Suppose this is not true. Then, by 8.5, we have

$$\{0\} = \text{null } T^0 \subsetneq \text{null } T^1 \subsetneq \cdots \subsetneq \text{null } T^{\dim V} \subsetneq \text{null } T^{\dim V+1},$$

where the symbol $\subsetneq$ means "contained in but not equal to". At each of the strict inclusions in the chain above, the dimension must increase by at least 1. Thus dim null $T^{\dim V+1} \geq \dim V + 1$, a contradiction because a subspace of $V$ cannot have a larger dimension than dim $V$.  ∎

Now we have the promised description of generalized eigenvectors.

*This corollary implies that the set of generalized eigenvectors of $T \in \mathcal{L}(V)$ corresponding to an eigenvalue $\lambda$ is a subspace of $V$.*

**8.7    Corollary:**  *Suppose $T \in \mathcal{L}(V)$ and $\lambda$ is an eigenvalue of $T$. Then the set of generalized eigenvectors of $T$ corresponding to $\lambda$ equals* null$(T - \lambda I)^{\dim V}$.

PROOF:  If $v \in \text{null}(T - \lambda I)^{\dim V}$, then clearly $v$ is a generalized eigenvector of $T$ corresponding to $\lambda$ (by the definition of generalized eigenvector).

Conversely, suppose that $v \in V$ is a generalized eigenvector of $T$ corresponding to $\lambda$. Thus there is a positive integer $j$ such that

$$v \in \text{null}(T - \lambda I)^j.$$

From 8.5 and 8.6 (with $T - \lambda I$ replacing $T$), we get $v \in \text{null}(T - \lambda I)^{\dim V}$, as desired.  ∎

An operator is called **nilpotent** if some power of it equals 0. For example, the operator $N \in \mathcal{L}(\mathbf{F}^4)$ defined by

$$N(z_1, z_2, z_3, z_4) = (z_3, z_4, 0, 0)$$

is nilpotent because $N^2 = 0$. As another example, the operator of differentiation on $\mathcal{P}_m(\mathbf{R})$ is nilpotent because the $(m+1)^{\text{st}}$ derivative of any polynomial of degree at most $m$ equals 0. Note that on this space of dimension $m + 1$, we need to raise the nilpotent operator to the power $m + 1$ to get 0. The next corollary shows that we never need to use a power higher than the dimension of the space.

*The Latin word **nil** means nothing or zero; the Latin word **potent** means power. Thus **nilpotent** literally means zero power.*

**8.8**    **Corollary:**  *Suppose $N \in \mathcal{L}(V)$ is nilpotent. Then $N^{\dim V} = 0$.*

PROOF:  Because $N$ is nilpotent, every vector in $V$ is a generalized eigenvector corresponding to the eigenvalue 0. Thus from 8.7 we see that null $N^{\dim V} = V$, as desired.                                             ∎

Having dealt with null spaces of powers of operators, we now turn our attention to ranges. Suppose $T \in \mathcal{L}(V)$ and $k$ is a nonnegative integer. If $w \in$ range $T^{k+1}$, then there exists $v \in V$ with

$$w = T^{k+1}v = T^k(Tv) \in \text{range } T^k.$$

Thus range $T^{k+1} \subset$ range $T^k$. In other words, we have

$$V = \text{range } T^0 \supset \text{range } T^1 \supset \cdots \supset \text{range } T^k \supset \text{range } T^{k+1} \supset \cdots .$$

*These inclusions go in the opposite direction from the corresponding inclusions for null spaces (8.4).*

The proposition below shows that the inclusions above become equalities once the power reaches the dimension of $V$.

**8.9**    **Proposition:**  *If $T \in \mathcal{L}(V)$, then*

$$\text{range } T^{\dim V} = \text{range } T^{\dim V + 1} = \text{range } T^{\dim V + 2} = \cdots .$$

PROOF:  We could prove this from scratch, but instead let's make use of the corresponding result already proved for null spaces. Suppose $m > \dim V$. Then

$$\begin{aligned}
\dim \text{range } T^m &= \dim V - \dim \text{null } T^m \\
&= \dim V - \dim \text{null } T^{\dim V} \\
&= \dim \text{range } T^{\dim V},
\end{aligned}$$

where the first and third equalities come from 3.4 and the second equality comes from 8.6. We already know that range $T^{\dim V} \supset$ range $T^m$. We just showed that dim range $T^{\dim V} =$ dim range $T^m$, so this implies that range $T^{\dim V} =$ range $T^m$, as desired. ∎

# The Characteristic Polynomial

Suppose $V$ is a complex vector space and $T \in \mathcal{L}(V)$. We know that $V$ has a basis with respect to which $T$ has an upper-triangular matrix (see 5.13). In general, this matrix is not unique—$V$ may have many different bases with respect to which $T$ has an upper-triangular matrix, and with respect to these different bases we may get different upper-triangular matrices. However, the diagonal of any such matrix must contain precisely the eigenvalues of $T$ (see 5.18). Thus if $T$ has dim $V$ distinct eigenvalues, then each one must appear exactly once on the diagonal of any upper-triangular matrix of $T$.

What if $T$ has fewer than dim $V$ distinct eigenvalues, as can easily happen? Then each eigenvalue must appear at least once on the diagonal of any upper-triangular matrix of $T$, but some of them must be repeated. Could the number of times that a particular eigenvalue is repeated depend on which basis of $V$ we choose?

You might guess that a number $\lambda$ appears on the diagonal of an upper-triangular matrix of $T$ precisely $\dim \text{null}(T - \lambda I)$ times. In general, this is false. For example, consider the operator on $\mathbf{C}^2$ whose matrix with respect to the standard basis is the upper-triangular matrix

$$\begin{bmatrix} 5 & 1 \\ 0 & 5 \end{bmatrix}.$$

*If $T$ happens to have a diagonal matrix $A$ with respect to some basis, then $\lambda$ appears on the diagonal of $A$ precisely $\dim \text{null}(T - \lambda I)$ times, as you should verify.*

For this operator, $\dim \text{null}(T - 5I) = 1$ but 5 appears on the diagonal twice. Note, however, that $\dim \text{null}(T - 5I)^2 = 2$ for this operator. This example illustrates the general situation—a number $\lambda$ appears on the diagonal of an upper-triangular matrix of $T$ precisely $\dim \text{null}(T - \lambda I)^{\dim V}$ times, as we will show in the following theorem. Because $\text{null}(T - \lambda I)^{\dim V}$ depends only on $T$ and $\lambda$ and not on a choice of basis, this implies that the number of times an eigenvalue is repeated on the diagonal of an upper-triangular matrix of $T$ is independent of which particular basis we choose. This result will be our key tool in analyzing the structure of an operator on a complex vector space.

**8.10    Theorem:**  *Let $T \in \mathcal{L}(V)$ and $\lambda \in \mathbf{F}$. Then for every basis of $V$ with respect to which $T$ has an upper-triangular matrix, $\lambda$ appears on the diagonal of the matrix of $T$ precisely $\dim \operatorname{null}(T - \lambda I)^{\dim V}$ times.*

PROOF:  We will assume, without loss of generality, that $\lambda = 0$ (once the theorem is proved in this case, the general case is obtained by replacing $T$ with $T - \lambda I$).

For convenience let $n = \dim V$. We will prove this theorem by induction on $n$. Clearly the desired result holds if $n = 1$. Thus we can assume that $n > 1$ and that the desired result holds on spaces of dimension $n - 1$.

Suppose $(v_1, \ldots, v_n)$ is a basis of $V$ with respect to which $T$ has an upper-triangular matrix

**8.11**
$$\begin{bmatrix} \lambda_1 & & & * \\ & \ddots & & \\ & & \lambda_{n-1} & \\ 0 & & & \lambda_n \end{bmatrix}.$$

*Recall that an asterisk is often used in matrices to denote entries that we do not know or care about.*

Let $U = \operatorname{span}(v_1, \ldots, v_{n-1})$. Clearly $U$ is invariant under $T$ (see 5.12), and the matrix of $T|_U$ with respect to the basis $(v_1, \ldots, v_{n-1})$ is

**8.12**
$$\begin{bmatrix} \lambda_1 & & * \\ & \ddots & \\ 0 & & \lambda_{n-1} \end{bmatrix}.$$

Thus, by our induction hypothesis, 0 appears on the diagonal of 8.12 $\dim \operatorname{null}(T|_U)^{n-1}$ times. We know that $\operatorname{null}(T|_U)^{n-1} = \operatorname{null}(T|_U)^n$ (because $U$ has dimension $n - 1$; see 8.6). Hence

**8.13**    0 appears on the diagonal of 8.12 $\dim \operatorname{null}(T|_U)^n$ times.

The proof breaks into two cases, depending on whether $\lambda_n = 0$. First consider the case where $\lambda_n \neq 0$. We will show that in this case

**8.14**                            $\operatorname{null} T^n \subset U.$

Once this has been verified, we will know that $\operatorname{null} T^n = \operatorname{null}(T|_U)^n$, and hence 8.13 will tell us that 0 appears on the diagonal of 8.11 exactly $\dim \operatorname{null} T^n$ times, completing the proof in the case where $\lambda_n \neq 0$.

Because $\mathcal{M}(T)$ is given by 8.11, we have

$$\mathcal{M}(T^n) = \mathcal{M}(T)^n = \begin{bmatrix} {\lambda_1}^n & & & * \\ & \ddots & & \\ & & {\lambda_{n-1}}^n & \\ 0 & & & {\lambda_n}^n \end{bmatrix}.$$

This shows that

$$T^n v_n = u + {\lambda_n}^n v_n$$

for some $u \in U$. To prove 8.14 (still assuming that $\lambda_n \neq 0$), suppose $v \in \operatorname{null} T^n$. We can write $v$ in the form

$$v = \tilde{u} + a v_n,$$

where $\tilde{u} \in U$ and $a \in \mathbf{F}$. Thus

$$0 = T^n v = T^n \tilde{u} + a T^n v_n = T^n \tilde{u} + a u + a {\lambda_n}^n v_n.$$

Because $T^n \tilde{u}$ and $au$ are in $U$ and $v_n \notin U$, this implies that $a{\lambda_n}^n = 0$. However, $\lambda_n \neq 0$, so $a = 0$. Thus $v = \tilde{u} \in U$, completing the proof of 8.14.

Now consider the case where $\lambda_n = 0$. In this case we will show that

**8.15**          $\dim \operatorname{null} T^n = \dim \operatorname{null}(T|_U)^n + 1,$

which along with 8.13 will complete the proof when $\lambda_n = 0$.

Using the formula for the dimension of the sum of two subspaces (2.18), we have

$$\dim \operatorname{null} T^n = \dim(U \cap \operatorname{null} T^n) + \dim(U + \operatorname{null} T^n) - \dim U$$
$$= \dim \operatorname{null}(T|_U)^n + \dim(U + \operatorname{null} T^n) - (n-1).$$

Suppose we can prove that $\operatorname{null} T^n$ contains a vector not in $U$. Then

$$n = \dim V \geq \dim(U + \operatorname{null} T^n) > \dim U = n - 1,$$

which implies that $\dim(U + \operatorname{null} T^n) = n$, which when combined with the formula above for $\dim \operatorname{null} T^n$ gives 8.15, as desired. Thus to complete the proof, we need only show that $\operatorname{null} T^n$ contains a vector not in $U$.

Let's think about how we might find a vector in $\operatorname{null} T^n$ that is not in $U$. We might try a vector of the form

$$u - v_n,$$

where $u \in U$. At least we are guaranteed that any such vector is not in $U$. Can we choose $u \in U$ such that the vector above is in null $T^n$? Let's compute:

$$T^n(u - v_n) = T^n u - T^n v_n.$$

To make the above vector equal 0, we must choose (if possible) $u \in U$ such that $T^n u = T^n v_n$. We can do this if $T^n v_n \in \text{range}(T|_U)^n$. Because 8.11 is the matrix of $T$ with respect to $(v_1, \ldots, v_n)$, we see that $T v_n \in U$ (recall that we are considering the case where $\lambda_n = 0$). Thus

$$T^n v_n = T^{n-1}(T v_n) \in \text{range}(T|_U)^{n-1} = \text{range}(T|_U)^n,$$

where the last equality comes from 8.9. In other words, we can indeed choose $u \in U$ such that $u - v_n \in$ null $T^n$, completing the proof.    ∎

Suppose $T \in \mathcal{L}(V)$. The **multiplicity** of an eigenvalue $\lambda$ of $T$ is defined to be the dimension of the subspace of generalized eigenvectors corresponding to $\lambda$. In other words, the multiplicity of an eigenvalue $\lambda$ of $T$ equals $\dim \text{null}(T - \lambda I)^{\dim V}$. If $T$ has an upper-triangular matrix with respect to some basis of $V$ (as always happens when $\mathbf{F} = \mathbf{C}$), then the multiplicity of $\lambda$ is simply the number of times $\lambda$ appears on the diagonal of this matrix (by the last theorem).

*Our definition of multiplicity has a clear connection with the geometric behavior of T. Most texts define multiplicity in terms of the multiplicity of the roots of a certain polynomial defined by determinants. These two definitions turn out to be equivalent.*

As an example of multiplicity, consider the operator $T \in \mathcal{L}(\mathbf{F}^3)$ defined by

**8.16**            $T(z_1, z_2, z_3) = (0, z_1, 5z_3).$

You should verify that 0 is an eigenvalue of $T$ with multiplicity 2, that 5 is an eigenvalue of $T$ with multiplicity 1, and that $T$ has no additional eigenvalues. As another example, if $T \in \mathcal{L}(\mathbf{F}^3)$ is the operator whose matrix is

**8.17**
$$\begin{bmatrix} 6 & 7 & 7 \\ 0 & 6 & 7 \\ 0 & 0 & 7 \end{bmatrix},$$

then 6 is an eigenvalue of $T$ with multiplicity 2 and 7 is an eigenvalue of $T$ with multiplicity 1 (this follows from the last theorem).

In each of the examples above, the sum of the multiplicities of the eigenvalues of $T$ equals 3, which is the dimension of the domain of $T$. The next proposition shows that this always happens on a complex vector space.

**8.18  Proposition:** *If $V$ is a complex vector space and $T \in \mathcal{L}(V)$, then the sum of the multiplicities of all the eigenvalues of $T$ equals $\dim V$.*

PROOF: Suppose $V$ is a complex vector space and $T \in \mathcal{L}(V)$. Then there is a basis of $V$ with respect to which the matrix of $T$ is upper triangular (by 5.13). The multiplicity of $\lambda$ equals the number of times $\lambda$ appears on the diagonal of this matrix (from 8.10). Because the diagonal of this matrix has length $\dim V$, the sum of the multiplicities of all the eigenvalues of $T$ must equal $\dim V$. ∎

Suppose $V$ is a complex vector space and $T \in \mathcal{L}(V)$. Let $\lambda_1, \ldots, \lambda_m$ denote the distinct eigenvalues of $T$. Let $d_j$ denote the multiplicity of $\lambda_j$ as an eigenvalue of $T$. The polynomial

$$(z - \lambda_1)^{d_1} \ldots (z - \lambda_m)^{d_m}$$

*Most texts define the characteristic polynomial using determinants. The approach taken here, which is considerably simpler, leads to an easy proof of the Cayley-Hamilton theorem.*

is called the ***characteristic polynomial*** of $T$. Note that the degree of the characteristic polynomial of $T$ equals $\dim V$ (from 8.18). Obviously the roots of the characteristic polynomial of $T$ equal the eigenvalues of $T$. As an example, the characteristic polynomial of the operator $T \in \mathcal{L}(\mathbf{C}^3)$ defined by 8.16 equals $z^2(z - 5)$.

Here is another description of the characteristic polynomial of an operator on a complex vector space. Suppose $V$ is a complex vector space and $T \in \mathcal{L}(V)$. Consider any basis of $V$ with respect to which $T$ has an upper-triangular matrix of the form

**8.19**
$$\mathcal{M}(T) = \begin{bmatrix} \lambda_1 & & * \\ & \ddots & \\ 0 & & \lambda_n \end{bmatrix}.$$

Then the characteristic polynomial of $T$ is given by

$$(z - \lambda_1) \ldots (z - \lambda_n);$$

this follows immediately from 8.10. As an example of this procedure, if $T \in \mathcal{L}(\mathbf{C}^3)$ is the operator whose matrix is given by 8.17, then the characteristic polynomial of $T$ equals $(z - 6)^2(z - 7)$.

In the next chapter we will define the characteristic polynomial of an operator on a real vector space and prove that the next result also holds for real vector spaces.

**8.20 Cayley-Hamilton Theorem:** *Suppose that $V$ is a complex vector space and $T \in \mathcal{L}(V)$. Let $q$ denote the characteristic polynomial of $T$. Then $q(T) = 0$.*

PROOF: Suppose $(v_1, \ldots, v_n)$ is a basis of $V$ with respect to which the matrix of $T$ has the upper-triangular form 8.19. To prove that $q(T) = 0$, we need only show that $q(T)v_j = 0$ for $j = 1, \ldots, n$. To do this, it suffices to show that

**8.21** $$(T - \lambda_1 I) \ldots (T - \lambda_j I)v_j = 0$$

for $j = 1, \ldots, n$.

We will prove 8.21 by induction on $j$. To get started, suppose $j = 1$. Because $\mathcal{M}(T, (v_1, \ldots, v_n))$ is given by 8.19, we have $Tv_1 = \lambda_1 v_1$, giving 8.21 when $j = 1$.

Now suppose that $1 < j \leq n$ and that

$$
\begin{aligned}
0 &= (T - \lambda_1 I)v_1 \\
&= (T - \lambda_1 I)(T - \lambda_2 I)v_2 \\
&\;\;\vdots \\
&= (T - \lambda_1 I) \ldots (T - \lambda_{j-1} I)v_{j-1}.
\end{aligned}
$$

Because $\mathcal{M}(T, (v_1, \ldots, v_n))$ is given by 8.19, we see that

$$(T - \lambda_j I)v_j \in \text{span}(v_1, \ldots, v_{j-1}).$$

Thus, by our induction hypothesis, $(T - \lambda_1 I) \ldots (T - \lambda_{j-1} I)$ applied to $(T - \lambda_j I)v_j$ gives 0. In other words, 8.21 holds, completing the proof. ∎

*The English mathematician Arthur Cayley published three mathematics papers before he completed his undergraduate degree in 1842. The Irish mathematician William Hamilton was made a professor in 1827 when he was 22 years old and still an undergraduate!*

## Decomposition of an Operator

We saw earlier that the domain of an operator might not decompose into invariant subspaces consisting of eigenvectors of the operator, even on a complex vector space. In this section we will see that every operator on a complex vector space has enough generalized eigenvectors to provide a decomposition.

We observed earlier that if $T \in \mathcal{L}(V)$, then null $T$ is invariant under $T$. Now we show that the null space of any polynomial of $T$ is also invariant under $T$.

**8.22**  **Proposition:**   *If $T \in \mathcal{L}(V)$ and $p \in \mathcal{P}(\mathbf{F})$, then* null $p(T)$ *is invariant under $T$.*

PROOF:  Suppose $T \in \mathcal{L}(V)$ and $p \in \mathcal{P}(\mathbf{F})$. Let $v \in$ null $p(T)$. Then $p(T)v = 0$. Thus

$$(p(T))(Tv) = T(p(T)v) = T(0) = 0,$$

and hence $Tv \in$ null $p(T)$. Thus null $p(T)$ is invariant under $T$, as desired.                                                                                                  ■

The following major structure theorem shows that every operator on a complex vector space can be thought of as composed of pieces, each of which is a nilpotent operator plus a scalar multiple of the identity. Actually we have already done all the hard work, so at this point the proof is easy.

**8.23**  **Theorem:**   *Suppose $V$ is a complex vector space and $T \in \mathcal{L}(V)$. Let $\lambda_1, \dots, \lambda_m$ be the distinct eigenvalues of $T$, and let $U_1, \dots, U_m$ be the corresponding subspaces of generalized eigenvectors. Then*

(a)      *$V = U_1 \oplus \cdots \oplus U_m$;*

(b)      *each $U_j$ is invariant under $T$;*

(c)      *each $(T - \lambda_j I)|_{U_j}$ is nilpotent.*

PROOF:  Note that $U_j = \text{null}(T - \lambda_j I)^{\dim V}$ for each $j$ (by 8.7). From 8.22 (with $p(z) = (z - \lambda_j)^{\dim V}$), we get (b). Obviously (c) follows from the definitions.

To prove (a), recall that the multiplicity of $\lambda_j$ as an eigenvalue of $T$ is defined to be $\dim U_j$ . The sum of these multiplicities equals $\dim V$ (see 8.18); thus

**8.24**                        $\dim V = \dim U_1 + \cdots + \dim U_m.$

Let $U = U_1 + \cdots + U_m$. Clearly $U$ is invariant under $T$. Thus we can define $S \in \mathcal{L}(U)$ by

$$S = T|_U.$$

Note that $S$ has the same eigenvalues, with the same multiplicities, as $T$ because all the generalized eigenvectors of $T$ are in $U$, the domain of $S$. Thus applying 8.18 to $S$, we get

$$\dim U = \dim U_1 + \cdots + \dim U_m.$$

This equation, along with 8.24, shows that $\dim V = \dim U$. Because $U$ is a subspace of $V$, this implies that $V = U$. In other words,

$$V = U_1 + \cdots + U_m.$$

This equation, along with 8.24, allows us to use 2.19 to conclude that (a) holds, completing the proof.                                        ∎

As we know, an operator on a complex vector space may not have enough eigenvectors to form a basis for the domain. The next result shows that on a complex vector space there are enough generalized eigenvectors to do this.

**8.25   Corollary:** *Suppose $V$ is a complex vector space and $T \in \mathcal{L}(V)$. Then there is a basis of $V$ consisting of generalized eigenvectors of $T$.*

PROOF:   Choose a basis for each $U_j$ in 8.23. Put all these bases together to form a basis of $V$ consisting of generalized eigenvectors of $T$.                                        ∎

Given an operator $T$ on $V$, we want to find a basis of $V$ so that the matrix of $T$ with respect to this basis is as simple as possible, meaning that the matrix contains many 0's. We begin by showing that if $N$ is nilpotent, we can choose a basis of $V$ such that the matrix of $N$ with respect to this basis has more than half of its entries equal to 0.

**8.26   Lemma:** *Suppose $N$ is a nilpotent operator on $V$. Then there is a basis of $V$ with respect to which the matrix of $N$ has the form*

**8.27**
$$\begin{bmatrix} 0 & & * \\ & \ddots & \\ 0 & & 0 \end{bmatrix};$$

*here all entries on and below the diagonal are 0's.*

PROOF:   First choose a basis of null $N$. Then extend this to a basis of null $N^2$. Then extend to a basis of null $N^3$. Continue in this fashion, eventually getting a basis of $V$ (because null $N^m = V$ for $m$ sufficiently large).

*If $V$ is complex vector space, a proof of this lemma follows easily from Exercise 6 in this chapter, 5.13, and 5.18. But the proof given here uses simpler ideas than needed to prove 5.13, and it works for both real and complex vector spaces.*

Now let's think about the matrix of $N$ with respect to this basis. The first column, and perhaps additional columns at the beginning, consists of all 0's because the corresponding basis vectors are in null $N$. The next set of columns comes from basis vectors in null $N^2$. Applying $N$ to any such vector, we get a vector in null $N$; in other words, we get a vector that is a linear combination of the previous basis vectors. Thus all nonzero entries in these columns must lie above the diagonal. The next set of columns come from basis vectors in null $N^3$. Applying $N$ to any such vector, we get a vector in null $N^2$; in other words, we get a vector that is a linear combination of the previous basis vectors. Thus, once again, all nonzero entries in these columns must lie above the diagonal. Continue in this fashion to complete the proof.                    ∎

Note that in the next theorem we get many more zeros in the matrix of $T$ than are needed to make it upper triangular.

**8.28**   **Theorem:** *Suppose $V$ is a complex vector space and $T \in \mathcal{L}(V)$. Let $\lambda_1, \ldots, \lambda_m$ be the distinct eigenvalues of $T$. Then there is a basis of $V$ with respect to which $T$ has a block diagonal matrix of the form*

$$\begin{bmatrix} A_1 & & 0 \\ & \ddots & \\ 0 & & A_m \end{bmatrix},$$

*where each $A_j$ is an upper-triangular matrix of the form*

**8.29**                    $$A_j = \begin{bmatrix} \lambda_j & & * \\ & \ddots & \\ 0 & & \lambda_j \end{bmatrix}.$$

PROOF:   For $j = 1, \ldots, m$, let $U_j$ denote the subspace of generalized eigenvectors of $T$ corresponding to $\lambda_j$. Thus $(T - \lambda_j I)|_{U_j}$ is nilpotent (see 8.23(c)). For each $j$, choose a basis of $U_j$ such that the matrix of $(T - \lambda_j I)|_{U_j}$ with respect to this basis is as in 8.26. Thus the matrix of $T|_{U_j}$ with respect to this basis will look like 8.29. Putting the bases for the $U_j$'s together gives a basis for $V$ (by 8.23(a)). The matrix of $T$ with respect to this basis has the desired form.                    ∎

# Square Roots

Recall that a square root of an operator $T \in \mathcal{L}(V)$ is an operator $S \in \mathcal{L}(V)$ such that $S^2 = T$. As an application of the main structure theorem from the last section, in this section we will show that every invertible operator on a complex vector space has a square root.

Every complex number has a square root, but not every operator on a complex vector space has a square root. An example of an operator on $\mathbf{C}^3$ that has no square root is given in Exercise 4 in this chapter. The noninvertibility of that particular operator is no accident, as we will soon see. We begin by showing that the identity plus a nilpotent operator always has a square root.

**8.30  Lemma:**  *Suppose $N \in \mathcal{L}(V)$ is nilpotent. Then $I + N$ has a square root.*

PROOF:  Consider the Taylor series for the function $\sqrt{1 + x}$:

**8.31**
$$\sqrt{1 + x} = 1 + a_1 x + a_2 x^2 + \cdots .$$

*Because $a_1 = 1/2$, this formula shows that $1 + x/2$ is a good estimate for $\sqrt{1 + x}$ when $x$ is small.*

We will not find an explicit formula for all the coefficients or worry about whether the infinite sum converges because we are using this equation only as motivation, not as a formal part of the proof.

Because $N$ is nilpotent, $N^m = 0$ for some positive integer $m$. In 8.31, suppose we replace $x$ with $N$ and 1 with $I$. Then the infinite sum on the right side becomes a finite sum (because $N^j = 0$ for all $j \geq m$). In other words, we guess that there is a square root of $I + N$ of the form

$$I + a_1 N + a_2 N^2 + \cdots + a_{m-1} N^{m-1}.$$

Having made this guess, we can try to choose $a_1, a_2, \ldots, a_{m-1}$ so that the operator above has its square equal to $I + N$. Now

$$\begin{aligned}
(I + &a_1 N + a_2 N^2 + a_3 N^3 + \cdots + a_{m-1} N^{m-1})^2 \\
&= I + 2a_1 N + (2a_2 + a_1{}^2)N^2 + (2a_3 + 2a_1 a_2)N^3 + \cdots \\
&\quad + (2a_{m-1} + \text{terms involving } a_1, \ldots, a_{m-2})N^{m-1}.
\end{aligned}$$

We want the right side of the equation above to equal $I + N$. Hence choose $a_1$ so that $2a_1 = 1$ (thus $a_1 = 1/2$). Next, choose $a_2$ so that $2a_2 + a_1{}^2 = 0$ (thus $a_2 = -1/8$). Then choose $a_3$ so that the coefficient of $N^3$ on the right side of the equation above equals 0 (thus $a_3 = 1/16$).

Continue in this fashion for $j = 4, \ldots, m - 1$, at each step solving for $a_j$ so that the coefficient of $N^j$ on the right side of the equation above equals 0. Actually we do not care about the explicit formula for the $a_j$'s. We need only know that some choice of the $a_j$'s gives a square root of $I + N$. ∎

The previous lemma is valid on real and complex vector spaces. However, the next result holds only on complex vector spaces.

*On real vector spaces there exist invertible operators that have no square roots. For example, the operator of multiplication by $-1$ on **R** has no square root because no real number has its square equal to $-1$.*

**8.32   Theorem:** *Suppose $V$ is a complex vector space. If $T \in \mathcal{L}(V)$ is invertible, then $T$ has a square root.*

PROOF: Suppose $T \in \mathcal{L}(V)$ is invertible. Let $\lambda_1, \ldots, \lambda_m$ be the distinct eigenvalues of $T$, and let $U_1, \ldots, U_m$ be the corresponding subspaces of generalized eigenvectors. For each $j$, there exists a nilpotent operator $N_j \in \mathcal{L}(U_j)$ such that $T|_{U_j} = \lambda_j I + N_j$ (see 8.23(c)). Because $T$ is invertible, none of the $\lambda_j$'s equals 0, so we can write

$$T|_{U_j} = \lambda_j \left( I + \frac{N_j}{\lambda_j} \right)$$

for each $j$. Clearly $N_j / \lambda_j$ is nilpotent, and so $I + N_j / \lambda_j$ has a square root (by 8.30). Multiplying a square root of the complex number $\lambda_j$ by a square root of $I + N_j / \lambda_j$, we obtain a square root $S_j$ of $T|_{U_j}$.

A typical vector $v \in V$ can be written uniquely in the form

$$v = u_1 + \cdots + u_m,$$

where each $u_j \in U_j$ (see 8.23). Using this decomposition, define an operator $S \in \mathcal{L}(V)$ by

$$Sv = S_1 u_1 + \cdots + S_m u_m.$$

You should verify that this operator $S$ is a square root of $T$, completing the proof. ∎

By imitating the techniques in this section, you should be able to prove that if $V$ is a complex vector space and $T \in \mathcal{L}(V)$ is invertible, then $T$ has a $k^{\text{th}}$-root for every positive integer $k$.

# The Minimal Polynomial

As we will soon see, given an operator on a finite-dimensional vector space, there is a unique monic polynomial of smallest degree that when applied to the operator gives 0. This polynomial is called the minimal polynomial of the operator and is the focus of attention in this section.

*A **monic polynomial** is a polynomial whose highest degree coefficient equals 1. For example, $2 + 3z^2 + z^8$ is a monic polynomial.*

Suppose $T \in \mathcal{L}(V)$, where $\dim V = n$. Then

$$(I, T, T^2, \ldots, T^{n^2})$$

cannot be linearly independent in $\mathcal{L}(V)$ because $\mathcal{L}(V)$ has dimension $n^2$ (see 3.20) and we have $n^2 + 1$ operators. Let $m$ be the smallest positive integer such that

**8.33**
$$(I, T, T^2, \ldots, T^m)$$

is linearly dependent. The linear dependence lemma (2.4) implies that one of the operators in the list above is a linear combination of the previous ones. Because $m$ was chosen to be the smallest positive integer such that 8.33 is linearly dependent, we conclude that $T^m$ is a linear combination of $(I, T, T^2, \ldots, T^{m-1})$. Thus there exist scalars $a_0, a_1, a_2, \ldots, a_{m-1} \in \mathbf{F}$ such that

$$a_0 I + a_1 T + a_2 T^2 + \cdots + a_{m-1} T^{m-1} + T^m = 0.$$

The choice of scalars $a_0, a_1, a_2, \ldots, a_{m-1} \in \mathbf{F}$ above is unique because two different such choices would contradict our choice of $m$ (subtracting two different equations of the form above, we would have a linearly dependent list shorter than 8.33). The polynomial

$$a_0 + a_1 z + a_2 z^2 + \cdots + a_{m-1} z^{m-1} + z^m$$

is called the **minimal polynomial** of $T$. It is the monic polynomial $p \in \mathcal{P}(\mathbf{F})$ of smallest degree such that $p(T) = 0$.

For example, the minimal polynomial of the identity operator $I$ is $z - 1$. The minimal polynomial of the operator on $\mathbf{F}^2$ whose matrix equals $\begin{bmatrix} 4 & 1 \\ 0 & 5 \end{bmatrix}$ is $20 - 9z + z^2$, as you should verify.

Clearly the degree of the minimal polynomial of each operator on $V$ is at most $(\dim V)^2$. The Cayley-Hamilton theorem (8.20) tells us that if $V$ is a complex vector space, then the minimal polynomial of each operator on $V$ has degree at most $\dim V$. This remarkable improvement also holds on real vector spaces, as we will see in the next chapter.

*Note that* $(z - \lambda)$
*divides a polynomial q*
*if and only if* $\lambda$ *is a*
*root of q. This follows*
*immediately from 4.1.*

A polynomial $p \in \mathcal{P}(\mathbf{F})$ is said to ***divide*** a polynomial $q \in \mathcal{P}(\mathbf{F})$ if there exists a polynomial $s \in \mathcal{P}(\mathbf{F})$ such that $q = sp$. In other words, $p$ divides $q$ if we can take the remainder $r$ in 4.6 to be 0. For example, the polynomial $(1 + 3z)^2$ divides $5 + 32z + 57z^2 + 18z^3$ because $5 + 32z + 57z^2 + 18z^3 = (2z + 5)(1 + 3z)^2$. Obviously every nonzero constant polynomial divides every polynomial.

The next result completely characterizes the polynomials that when applied to an operator give the 0 operator.

**8.34**   **Theorem:**  *Let* $T \in \mathcal{L}(V)$ *and let* $q \in \mathcal{P}(\mathbf{F})$. *Then* $q(T) = 0$ *if and only if the minimal polynomial of* $T$ *divides* $q$.

PROOF:  Let $p$ denote the minimal polynomial of $T$.

First we prove the easy direction. Suppose that $p$ divides $q$. Thus there exists a polynomial $s \in \mathcal{P}(\mathbf{F})$ such that $q = sp$. We have

$$q(T) = s(T)p(T) = s(T)0 = 0,$$

as desired.

To prove the other direction, suppose that $q(T) = 0$. By the division algorithm (4.5), there exist polynomials $s, r \in \mathcal{P}(\mathbf{F})$ such that

**8.35**                                     $q = sp + r$

and $\deg r < \deg p$. We have

$$0 = q(T) = s(T)p(T) + r(T) = r(T).$$

Because $p$ is the minimal polynomial of $T$ and $\deg r < \deg p$, the equation above implies that $r = 0$. Thus 8.35 becomes the equation $q = sp$, and hence $p$ divides $q$, as desired.          ∎

Now we describe the eigenvalues of an operator in terms of its minimal polynomial.

**8.36**   **Theorem:**  *Let* $T \in \mathcal{L}(V)$. *Then the roots of the minimal polynomial of* $T$ *are precisely the eigenvalues of* $T$.

PROOF: Let

$$p(z) = a_0 + a_1 z + a_2 z^2 + \cdots + a_{m-1} z^{m-1} + z^m$$

be the minimal polynomial of $T$.

First suppose that $\lambda \in \mathbf{F}$ is a root of $p$. Then the minimal polynomial of $T$ can be written in the form

$$p(z) = (z - \lambda)q(z),$$

where $q$ is a monic polynomial with coefficients in $\mathbf{F}$ (see 4.1). Because $p(T) = 0$, we have

$$0 = (T - \lambda I)(q(T)v)$$

for all $v \in V$. Because the degree of $q$ is less than the degree of the minimal polynomial $p$, there must exist at least one vector $v \in V$ such that $q(T)v \neq 0$. The equation above thus implies that $\lambda$ is an eigenvalue of $T$, as desired.

To prove the other direction, now suppose that $\lambda \in \mathbf{F}$ is an eigenvalue of $T$. Let $v$ be a nonzero vector in $V$ such that $Tv = \lambda v$. Repeated applications of $T$ to both sides of this equation show that $T^j v = \lambda^j v$ for every nonnegative integer $j$. Thus

$$\begin{aligned} 0 = p(T)v &= (a_0 + a_1 T + a_2 T^2 + \cdots + a_{m-1} T^{m-1} + T^m)v \\ &= (a_0 + a_1 \lambda + a_2 \lambda^2 + \cdots + a_{m-1} \lambda^{m-1} + \lambda^m)v \\ &= p(\lambda)v. \end{aligned}$$

Because $v \neq 0$, the equation above implies that $p(\lambda) = 0$, as desired. ∎

Suppose we are given, in concrete form, the matrix (with respect to some basis) of some operator $T \in \mathcal{L}(V)$. To find the minimal polynomial of $T$, consider

$$(\mathcal{M}(I), \mathcal{M}(T), \mathcal{M}(T)^2, \ldots, \mathcal{M}(T)^m)$$

for $m = 1, 2, \ldots$ until this list is linearly dependent. Then find the scalars $a_0, a_1, a_2, \ldots, a_{m-1} \in \mathbf{F}$ such that

$$a_0 \mathcal{M}(I) + a_1 \mathcal{M}(T) + a_2 \mathcal{M}(T)^2 + \cdots + a_{m-1} \mathcal{M}(T)^{m-1} + \mathcal{M}(T)^m = 0.$$

*You can think of this as a system of $(\dim V)^2$ equations in $m$ variables $a_0, a_1, \ldots, a_{m-1}$.*

The scalars $a_0, a_1, a_2, \ldots, a_{m-1}, 1$ will then be the coefficients of the minimal polynomial of $T$. All this can be computed using a familiar process such as Gaussian elimination.

For example, consider the operator $T$ on $\mathbb{C}^5$ whose matrix is given by

**8.37**
$$\begin{bmatrix} 0 & 0 & 0 & 0 & -3 \\ 1 & 0 & 0 & 0 & 6 \\ 0 & 1 & 0 & 0 & 0 \\ 0 & 0 & 1 & 0 & 0 \\ 0 & 0 & 0 & 1 & 0 \end{bmatrix}.$$

Because of the large number of 0's in this matrix, Gaussian elimination is not needed here. Simply compute powers of $\mathcal{M}(T)$ and notice that there is no linear dependence until the fifth power. Do the computations and you will see that the minimal polynomial of $T$ equals

**8.38**
$$z^5 - 6z + 3.$$

Now what about the eigenvalues of this particular operator? From 8.36, we see that the eigenvalues of $T$ equal the solutions to the equation

$$z^5 - 6z + 3 = 0.$$

Unfortunately no solution to this equation can be computed using rational numbers, arbitrary roots of rational numbers, and the usual rules of arithmetic (a proof of this would take us considerably beyond linear algebra). Thus we cannot find an exact expression for any eigenvalues of $T$ in any familiar form, though numeric techniques can give good approximations for the eigenvalues of $T$. The numeric techniques, which we will not discuss here, show that the eigenvalues for this particular operator are approximately

$$-1.67, \quad 0.51, \quad 1.40, \quad -0.12 + 1.59i, \quad -0.12 - 1.59i.$$

Note that the nonreal eigenvalues occur as a pair, with each the complex conjugate of the other, as expected for the roots of a polynomial with real coefficients (see 4.10).

Suppose $V$ is a complex vector space and $T \in \mathcal{L}(V)$. The Cayley-Hamilton theorem (8.20) and 8.34 imply that the minimal polynomial of $T$ divides the characteristic polynomial of $T$. Both these polynomials are monic. Thus if the minimal polynomial of $T$ has degree $\dim V$, then it must equal the characteristic polynomial of $T$. For example, if $T$ is the operator on $\mathbb{C}^5$ whose matrix is given by 8.37, then the characteristic polynomial of $T$, as well as the minimal polynomial of $T$, is given by 8.38.

# Jordan Form

We know that if $V$ is a complex vector space, then for every $T \in \mathcal{L}(V)$ there is a basis of $V$ with respect to which $T$ has a nice upper-triangular matrix (see 8.28). In this section we will see that we can do even better—there is a basis of $V$ with respect to which the matrix of $T$ contains zeros everywhere except possibly on the diagonal and the line directly above the diagonal.

We begin by describing the nilpotent operators. Consider, for example, the nilpotent operator $N \in \mathcal{L}(\mathbf{F}^n)$ defined by

$$N(z_1, \ldots, z_n) = (0, z_1, \ldots, z_{n-1}).$$

If $v = (1, 0, \ldots, 0)$, then clearly $(v, Nv, \ldots, N^{n-1}v)$ is a basis of $\mathbf{F}^n$ and $(N^{n-1}v)$ is a basis of null $N$, which has dimension 1.

As another example, consider the nilpotent operator $N \in \mathcal{L}(\mathbf{F}^5)$ defined by

**8.39**      $$N(z_1, z_2, z_3, z_4, z_5) = (0, z_1, z_2, 0, z_4).$$

Unlike the nilpotent operator discussed in the previous paragraph, for this nilpotent operator there does not exist a vector $v \in \mathbf{F}^5$ such that $(v, Nv, N^2v, N^3v, N^4v)$ is a basis of $\mathbf{F}^5$. However, if $v_1 = (1, 0, 0, 0, 0)$ and $v_2 = (0, 0, 0, 1, 0)$, then $(v_1, Nv_1, N^2v_1, v_2, Nv_2)$ is a basis of $\mathbf{F}^5$ and $(N^2v_1, Nv_2)$ is a basis of null $N$, which has dimension 2.

Suppose $N \in \mathcal{L}(V)$ is nilpotent. For each nonzero vector $v \in V$, let $m(v)$ denote the largest nonnegative integer such that $N^{m(v)}v \neq 0$. For example, if $N \in \mathcal{L}(\mathbf{F}^5)$ is defined by 8.39, then $m(1, 0, 0, 0, 0) = 2$.

The lemma below shows that every nilpotent operator $N \in \mathcal{L}(V)$ behaves similarly to the example defined by 8.39, in the sense that there is a finite collection of vectors $v_1, \ldots, v_k \in V$ such that the nonzero vectors of the form $N^j v_r$ form a basis of $V$; here $r$ varies from 1 to $k$ and $j$ varies from 0 to $m(v_r)$.

*Obviously $m(v)$ depends on $N$ as well as on $v$, but the choice of $N$ will be clear from the context.*

**8.40   Lemma:**   *If $N \in \mathcal{L}(V)$ is nilpotent, then there exist vectors $v_1, \ldots, v_k \in V$ such that*

(a)   $(v_1, Nv_1, \ldots, N^{m(v_1)}v_1, \ldots, v_k, Nv_k, \ldots, N^{m(v_k)}v_k)$ *is a basis of $V$;*

(b)   $(N^{m(v_1)}v_1, \ldots, N^{m(v_k)}v_k)$ *is a basis of* null $N$.

PROOF: Suppose $N$ is nilpotent. Then $N$ is not injective and thus $\dim \operatorname{range} N < \dim V$ (see 3.21). By induction on the dimension of $V$, we can assume that the lemma holds on all vector spaces of smaller dimension. Using $\operatorname{range} N$ in place of $V$ and $N|_{\operatorname{range} N}$ in place of $N$, we thus have vectors $u_1, \ldots, u_j \in \operatorname{range} N$ such that

(i)     $(u_1, Nu_1, \ldots, N^{m(u_1)}u_1, \ldots, u_j, Nu_j, \ldots, N^{m(u_j)}u_j)$ is a basis of $\operatorname{range} N$;

(ii)    $(N^{m(u_1)}u_1, \ldots, N^{m(u_j)}u_j)$ is a basis of $\operatorname{null} N \cap \operatorname{range} N$.

Because each $u_r \in \operatorname{range} N$, we can choose $v_1, \ldots, v_j \in V$ such that $Nv_r = u_r$ for each $r$. Note that $m(v_r) = m(u_r) + 1$ for each $r$.

*The existence of a subspace W with this property follows from 2.13.*

Let $W$ be a subspace of $\operatorname{null} N$ such that

**8.41**                    $\operatorname{null} N = (\operatorname{null} N \cap \operatorname{range} N) \oplus W$

and choose a basis of $W$, which we will label $(v_{j+1}, \ldots, v_k)$. Because $v_{j+1}, \ldots, v_k \in \operatorname{null} N$, we have $m(v_{j+1}) = \cdots = m(v_k) = 0$.

Having constructed $v_1, \ldots, v_k$, we now need to show that (a) and (b) hold. We begin by showing that the alleged basis in (a) is linearly independent. To do this, suppose

**8.42**                    $$0 = \sum_{r=1}^{k} \sum_{s=0}^{m(v_r)} a_{r,s} N^s(v_r),$$

where each $a_{r,s} \in \mathbf{F}$. Applying $N$ to both sides of the equation above, we get

$$0 = \sum_{r=1}^{k} \sum_{s=0}^{m(v_r)} a_{r,s} N^{s+1}(v_r)$$

$$= \sum_{r=1}^{j} \sum_{s=0}^{m(u_r)} a_{r,s} N^s(u_r).$$

The last equation, along with (i), implies that $a_{r,s} = 0$ for $1 \le r \le j$, $0 \le s \le m(v_r) - 1$. Thus 8.42 reduces to the equation

$$0 = a_{1,m(v_1)} N^{m(v_1)} v_1 + \cdots + a_{j,m(v_j)} N^{m(v_j)} v_j$$

$$+ a_{j+1,0} v_{j+1} + \cdots + a_{k,0} v_k.$$

The terms on the first line on the right are all in null $N \cap$ range $N$; the terms on the second line are all in $W$. Thus the last equation and 8.41 imply that

$$0 = a_{1,m(v_1)}N^{m(v_1)}v_1 + \cdots + a_{j,m(v_j)}N^{m(v_j)}v_j$$

**8.43**
$$= a_{1,m(v_1)}N^{m(u_1)}u_1 + \cdots + a_{j,m(v_j)}N^{m(u_j)}u_j$$

and

**8.44**
$$0 = a_{j+1,0}v_{j+1} + \cdots + a_{k,0}v_k.$$

Now 8.43 and (ii) imply that $a_{1,m(v_1)} = \cdots = a_{j,m(v_j)} = 0$. Because $(v_{j+1}, \ldots, v_k)$ is a basis of $W$, 8.44 implies that $a_{j+1,0} = \cdots = a_{k,0} = 0$. Thus all the $a$'s equal 0, and hence the list of vectors in (a) is linearly independent.

Clearly (ii) implies that dim(null $N \cap$ range $N$) $= j$. Along with 8.41, this implies that

**8.45**
$$\dim \text{null } N = k.$$

Clearly (i) implies that

$$\dim \text{range } N = \sum_{r=0}^{j} (m(u_r) + 1)$$

**8.46**
$$= \sum_{r=0}^{j} m(v_r).$$

The list of vectors in (a) has length

$$\sum_{r=0}^{k} (m(v_r) + 1) = k + \sum_{r=0}^{j} m(v_r)$$
$$= \dim \text{null } N + \dim \text{range } N$$
$$= \dim V,$$

where the second equality comes from 8.45 and 8.46, and the third equality comes from 3.4. The last equation shows that the list of vectors in (a) has length dim $V$; because this list is linearly independent, it is a basis of $V$ (see 2.17), completing the proof of (a).

Finally, note that

$$(N^{m(v_1)}v_1, \ldots, N^{m(v_k)}v_k) = (N^{m(u_1)}u_1, \ldots, N^{m(u_j)}u_j, v_{j+1}, \ldots, v_k).$$

LIVERPOOL
JOHN MOORES UNIVERSITY
AVRIL ROBARTS LRC
TEL. 0151 231 4022

Now (ii) and 8.41 show that the last list above is a basis of null $N$, completing the proof of (b).　　　　　　■

Suppose $T \in \mathcal{L}(V)$. A basis of $V$ is called a **Jordan basis** for $T$ if with respect to this basis $T$ has a block diagonal matrix

$$
\begin{bmatrix}
A_1 & & 0 \\
& \ddots & \\
0 & & A_m
\end{bmatrix},
$$

where each $A_j$ is an upper-triangular matrix of the form

$$
A_j =
\begin{bmatrix}
\lambda_j & 1 & & 0 \\
& \ddots & \ddots & \\
& & \ddots & 1 \\
0 & & & \lambda_j
\end{bmatrix}.
$$

*To understand why each $\lambda_j$ must be an eigenvalue of $T$, see 5.18.*

In each $A_j$, the diagonal is filled with some eigenvalue $\lambda_j$ of $T$, the line directly above the diagonal is filled with 1's, and all other entries are 0 ($A_j$ may be just a 1-by-1 block consisting of just some eigenvalue).

Because there exist operators on real vector spaces that have no eigenvalues, there exist operators on real vector spaces for which there is no corresponding Jordan basis. Thus the hypothesis that $V$ is a complex vector space is required for the next result, even though the previous lemma holds on both real and complex vector spaces.

*The French mathematician Camille Jordan first published a proof of this theorem in 1870.*

**8.47　Theorem:** *Suppose $V$ is a complex vector space. If $T \in \mathcal{L}(V)$, then there is a basis of $V$ that is a Jordan basis for $T$.*

PROOF: First consider a nilpotent operator $N \in \mathcal{L}(V)$ and the vectors $v_1, \ldots, v_k \in V$ given by 8.40. For each $j$, note that $N$ sends the first vector in the list $(N^{m(v_j)} v_j, \ldots, N v_j, v_j)$ to 0 and that $N$ sends each vector in this list other than the first vector to the previous vector. In other words, if we reverse the order of the basis given by 8.40(a), then we obtain a basis of $V$ with respect to which $N$ has a block diagonal matrix, where each matrix on the diagonal has the form

$$
\begin{bmatrix}
0 & 1 & & 0 \\
& \ddots & \ddots & \\
& & \ddots & 1 \\
0 & & & 0
\end{bmatrix}.
$$

Thus the theorem holds for nilpotent operators.

Now suppose $T \in \mathcal{L}(V)$. Let $\lambda_1, \ldots, \lambda_m$ be the distinct eigenvalues of $T$, with $U_1, \ldots, U_m$ the corresponding subspaces of generalized eigenvectors. We have

$$V = U_1 \oplus \cdots \oplus U_m,$$

where each $(T - \lambda_j I)|_{U_j}$ is nilpotent (see 8.23). By the previous paragraph, there is a basis of each $U_j$ that is a Jordan basis for $(T - \lambda_j I)|_{U_j}$. Putting these bases together gives a basis of $V$ that is a Jordan basis for $T$. ∎

## *Exercises*

1.   Define $T \in \mathcal{L}(\mathbf{C}^2)$ by

$$T(w, z) = (z, 0).$$

Find all generalized eigenvectors of $T$.

2.   Define $T \in \mathcal{L}(\mathbf{C}^2)$ by

$$T(w, z) = (-z, w).$$

Find all generalized eigenvectors of $T$.

3.   Suppose $T \in \mathcal{L}(V)$, $m$ is a positive integer, and $v \in V$ is such that $T^{m-1}v \neq 0$ but $T^m v = 0$. Prove that

$$(v, Tv, T^2 v, \ldots, T^{m-1}v)$$

is linearly independent.

4.   Suppose $T \in \mathcal{L}(\mathbf{C}^3)$ is defined by $T(z_1, z_2, z_3) = (z_2, z_3, 0)$. Prove that $T$ has no square root. More precisely, prove that there does not exist $S \in \mathcal{L}(\mathbf{C}^3)$ such that $S^2 = T$.

5.   Suppose $S, T \in \mathcal{L}(V)$. Prove that if $ST$ is nilpotent, then $TS$ is nilpotent.

6.   Suppose $N \in \mathcal{L}(V)$ is nilpotent. Prove (without using 8.26) that $0$ is the only eigenvalue of $N$.

7.   Suppose $V$ is an inner-product space. Prove that if $N \in \mathcal{L}(V)$ is self-adjoint and nilpotent, then $N = 0$.

8.   Suppose $N \in \mathcal{L}(V)$ is such that $\mathrm{null}\, N^{\dim V - 1} \neq \mathrm{null}\, N^{\dim V}$. Prove that $N$ is nilpotent and that

$$\dim \mathrm{null}\, N^j = j$$

for every integer $j$ with $0 \leq j \leq \dim V$.

9.   Suppose $T \in \mathcal{L}(V)$ and $m$ is a nonnegative integer such that

$$\mathrm{range}\, T^m = \mathrm{range}\, T^{m+1}.$$

Prove that $\mathrm{range}\, T^k = \mathrm{range}\, T^m$ for all $k > m$.

10. Prove or give a counterexample: if $T \in \mathcal{L}(V)$, then

$$V = \text{null } T \oplus \text{range } T.$$

11. Prove that if $T \in \mathcal{L}(V)$, then

$$V = \text{null } T^n \oplus \text{range } T^n,$$

where $n = \dim V$.

12. Suppose $V$ is a complex vector space, $N \in \mathcal{L}(V)$, and 0 is the only eigenvalue of $N$. Prove that $N$ is nilpotent. Give an example to show that this is not necessarily true on a real vector space.

13. Suppose that $V$ is a complex vector space with $\dim V = n$ and $T \in \mathcal{L}(V)$ is such that

$$\text{null } T^{n-2} \neq \text{null } T^{n-1}.$$

Prove that $T$ has at most two distinct eigenvalues.

14. Give an example of an operator on $\mathbf{C}^4$ whose characteristic polynomial equals $(z - 7)^2(z - 8)^2$.

15. Suppose $V$ is a complex vector space. Suppose $T \in \mathcal{L}(V)$ is such that 5 and 6 are eigenvalues of $T$ and that $T$ has no other eigenvalues. Prove that

$$(T - 5I)^{n-1}(T - 6I)^{n-1} = 0,$$

where $n = \dim V$.

16. Suppose $V$ is a complex vector space and $T \in \mathcal{L}(V)$. Prove that $V$ has a basis consisting of eigenvectors of $T$ if and only if every generalized eigenvector of $T$ is an eigenvector of $T$.

*For complex vector spaces, this exercise adds another equivalence to the list given by 5.21.*

17. Suppose $V$ is an inner-product space and $N \in \mathcal{L}(V)$ is nilpotent. Prove that there exists an orthonormal basis of $V$ with respect to which $N$ has an upper-triangular matrix.

18. Define $N \in \mathcal{L}(\mathbf{F}^5)$ by

$$N(x_1, x_2, x_3, x_4, x_5) = (2x_2, 3x_3, -x_4, 4x_5, 0).$$

Find a square root of $I + N$.

19.  Prove that if $V$ is a complex vector space, then every invertible operator on $V$ has a cube root.

20.  Suppose $T \in \mathcal{L}(V)$ is invertible. Prove that there exists a polynomial $p \in \mathcal{P}(\mathbf{F})$ such that $T^{-1} = p(T)$.

21.  Give an example of an operator on $\mathbf{C}^3$ whose minimal polynomial equals $z^2$.

22.  Give an example of an operator on $\mathbf{C}^4$ whose minimal polynomial equals $z(z-1)^2$.

*For complex vector*
*spaces, this exercise*
*adds another*
*equivalence to the list*
*given by 5.21.*

23.  Suppose $V$ is a complex vector space and $T \in \mathcal{L}(V)$. Prove that $V$ has a basis consisting of eigenvectors of $T$ if and only if the minimal polynomial of $T$ has no repeated roots.

24.  Suppose $V$ is an inner-product space. Prove that if $T \in \mathcal{L}(V)$ is normal, then the minimal polynomial of $T$ has no repeated roots.

25.  Suppose $T \in \mathcal{L}(V)$ and $v \in V$. Let $p$ be the monic polynomial of smallest degree such that

$$p(T)v = 0.$$

Prove that $p$ divides the minimal polynomial of $T$.

26.  Give an example of an operator on $\mathbf{C}^4$ whose characteristic and minimal polynomials both equal $z(z-1)^2(z-3)$.

27.  Give an example of an operator on $\mathbf{C}^4$ whose characteristic polynomial equals $z(z-1)^2(z-3)$ and whose minimal polynomial equals $z(z-1)(z-3)$.

*This exercise shows*
*that every monic*
*polynomial is the*
*characteristic*
*polynomial of some*
*operator.*

28.  Suppose $a_0, \ldots, a_{n-1} \in \mathbf{C}$. Find the minimal and characteristic polynomials of the operator on $\mathbf{C}^n$ whose matrix (with respect to the standard basis) is

$$\begin{bmatrix} 0 & & & & & -a_0 \\ 1 & 0 & & & & -a_1 \\ & 1 & \ddots & & & -a_2 \\ & & \ddots & \ddots & & \vdots \\ & & & & 0 & -a_{n-2} \\ & & & & 1 & -a_{n-1} \end{bmatrix}.$$

29.    Suppose $N \in \mathcal{L}(V)$ is nilpotent. Prove that the minimal poly-
       nomial of $N$ is $z^{m+1}$, where $m$ is the length of the longest con-
       secutive string of 1's that appears on the line directly above the
       diagonal in the matrix of $N$ with respect to any Jordan basis for $N$.

30.    Suppose $V$ is a complex vector space and $T \in \mathcal{L}(V)$. Prove that
       there does not exist a direct sum decomposition of $V$ into two
       proper subspaces invariant under $T$ if and only if the minimal
       polynomial of $T$ is of the form $(z - \lambda)^{\dim V}$ for some $\lambda \in \mathbf{C}$.

31.    Suppose $T \in \mathcal{L}(V)$ and $(v_1, \ldots, v_n)$ is a basis of $V$ that is a Jordan
       basis for $T$. Describe the matrix of $T$ with respect to the basis
       $(v_n, \ldots, v_1)$ obtained by reversing the order of the $v$'s.

# CHAPTER 9

# *Operators on Real Vector Spaces*

In this chapter we delve deeper into the structure of operators on real vector spaces. The important results here are somewhat more complex than the analogous results from the last chapter on complex vector spaces.

> Recall that **F** denotes **R** or **C**.
> Also, $V$ is a finite-dimensional, nonzero vector space over **F**.

Some of the new results in this chapter are valid on complex vector spaces, so we have not assumed that $V$ is a real vector space.

# *Eigenvalues of Square Matrices*

We have defined eigenvalues of operators; now we need to extend that notion to square matrices. Suppose $A$ is an $n$-by-$n$ matrix with entries in $\mathbf{F}$. A number $\lambda \in \mathbf{F}$ is called an *eigenvalue* of $A$ if there exists a nonzero $n$-by-1 matrix $x$ such that

$$Ax = \lambda x.$$

For example, 3 is an eigenvalue of $\left[\begin{smallmatrix} 7 & 8 \\ 1 & 5 \end{smallmatrix}\right]$ because

$$\begin{bmatrix} 7 & 8 \\ 1 & 5 \end{bmatrix} \begin{bmatrix} 2 \\ -1 \end{bmatrix} = \begin{bmatrix} 6 \\ -3 \end{bmatrix} = 3 \begin{bmatrix} 2 \\ -1 \end{bmatrix}.$$

As another example, you should verify that the matrix $\left[\begin{smallmatrix} 0 & -1 \\ 1 & 0 \end{smallmatrix}\right]$ has no eigenvalues if we are thinking of $\mathbf{F}$ as the real numbers (by definition, an eigenvalue must be in $\mathbf{F}$) and has eigenvalues $i$ and $-i$ if we are thinking of $\mathbf{F}$ as the complex numbers.

We now have two notions of eigenvalue—one for operators and one for square matrices. As you might expect, these two notions are closely connected, as we now show.

**9.1**    **Proposition:**  *Suppose $T \in \mathcal{L}(V)$ and $A$ is the matrix of $T$ with respect to some basis of $V$. Then the eigenvalues of $T$ are the same as the eigenvalues of $A$.*

PROOF:   Let $(v_1, \ldots, v_n)$ be the basis of $V$ with respect to which $T$ has matrix $A$. Let $\lambda \in \mathbf{F}$. We need to show that $\lambda$ is an eigenvalue of $T$ if and only if $\lambda$ is an eigenvalue of $A$.

First suppose $\lambda$ is an eigenvalue of $T$. Let $v \in V$ be a nonzero vector such that $Tv = \lambda v$. We can write

**9.2**                    $$v = a_1 v_1 + \cdots + a_n v_n,$$

where $a_1, \ldots, a_n \in \mathbf{F}$. Let $x$ be the matrix of the vector $v$ with respect to the basis $(v_1, \ldots, v_n)$. Recall from Chapter 3 that this means

**9.3**                    $$x = \begin{bmatrix} a_1 \\ \vdots \\ a_n \end{bmatrix}.$$

We have

$$Ax = \mathcal{M}(T)\mathcal{M}(v) = \mathcal{M}(Tv) = \mathcal{M}(\lambda v) = \lambda \mathcal{M}(v) = \lambda x,$$

where the second equality comes from 3.14. The equation above shows that $\lambda$ is an eigenvalue of $A$, as desired.

To prove the implication in the other direction, now suppose $\lambda$ is an eigenvalue of $A$. Let $x$ be a nonzero $n$-by-1 matrix such that $Ax = \lambda x$. We can write $x$ in the form 9.3 for some scalars $a_1, \ldots, a_n \in \mathbf{F}$. Define $v \in V$ by 9.2. Then

$$\mathcal{M}(Tv) = \mathcal{M}(T)\mathcal{M}(v) = Ax = \lambda x = \mathcal{M}(\lambda v).$$

where the first equality comes from 3.14. The equation above implies that $Tv = \lambda v$, and thus $\lambda$ is an eigenvalue of $T$, completing the proof. ■

Because every square matrix is the matrix of some operator, the proposition above allows us to translate results about eigenvalues of operators into the language of eigenvalues of square matrices. For example, every square matrix of complex numbers has an eigenvalue (from 5.10). As another example, every $n$-by-$n$ matrix has at most $n$ distinct eigenvalues (from 5.9).

## Block Upper-Triangular Matrices

Earlier we proved that each operator on a complex vector space has an upper-triangular matrix with respect to some basis (see 5.13). In this section we will see that we can almost do as well on real vector spaces.

In the last two chapters we used block diagonal matrices, which extend the notion of diagonal matrices. Now we will need to use the corresponding extension of upper-triangular matrices. A **block upper-triangular matrix** is a square matrix of the form

$$\begin{bmatrix} A_1 & & * \\ & \ddots & \\ 0 & & A_m \end{bmatrix},$$

*As usual, we use an asterisk to denote entries of the matrix that play no important role in the topics under consideration.*

where $A_1, \ldots, A_m$ are square matrices lying along the diagonal, all entries below $A_1, \ldots, A_m$ equal 0, and the $*$ denotes arbitrary entries. For example, the matrix

$$A = \begin{bmatrix} 4 & 10 & 11 & 12 & 13 \\ 0 & -3 & -3 & 14 & 25 \\ 0 & -3 & -3 & 16 & 17 \\ 0 & 0 & 0 & 5 & 5 \\ 0 & 0 & 0 & 5 & 5 \end{bmatrix}$$

is a block upper-triangular matrix with

$$A = \begin{bmatrix} A_1 & & * \\ & A_2 & \\ 0 & & A_3 \end{bmatrix},$$

where

$$A_1 = \begin{bmatrix} 4 \end{bmatrix}, \quad A_2 = \begin{bmatrix} -3 & -3 \\ -3 & -3 \end{bmatrix}, \quad A_3 = \begin{bmatrix} 5 & 5 \\ 5 & 5 \end{bmatrix}.$$

*Every upper-triangular matrix is also a block upper-triangular matrix with blocks of size 1-by-1 along the diagonal. At the other extreme, every square matrix is a block upper-triangular matrix because we can take the first (and only) block to be the entire matrix. Smaller blocks are better in the sense that the matrix then has more 0's.*

Now we prove that for each operator on a real vector space, we can find a basis that gives a block upper-triangular matrix with blocks of size at most 2-by-2 on the diagonal.

**9.4    Theorem:** *Suppose $V$ is a real vector space and $T \in \mathcal{L}(V)$. Then there is a basis of $V$ with respect to which $T$ has a block upper-triangular matrix*

**9.5**
$$\begin{bmatrix} A_1 & & * \\ & \ddots & \\ 0 & & A_m \end{bmatrix},$$

*where each $A_j$ is a 1-by-1 matrix or a 2-by-2 matrix with no eigenvalues.*

PROOF:  Clearly the desired result holds if $\dim V = 1$.

Next, consider the case where $\dim V = 2$. If $T$ has an eigenvalue $\lambda$, then let $v_1 \in V$ be any nonzero eigenvector. Extend $(v_1)$ to a basis $(v_1, v_2)$ of $V$. With respect to this basis, $T$ has an upper-triangular matrix of the form

$$\begin{bmatrix} \lambda & a \\ 0 & b \end{bmatrix}.$$

In particular, if $T$ has an eigenvalue, then there is a basis of $V$ with respect to which $T$ has an upper-triangular matrix. If $T$ has no eigenvalues, then choose any basis $(v_1, v_2)$ of $V$. With respect to this basis, the matrix of $T$ has no eigenvalues (by 9.1). Thus regardless of whether $T$ has eigenvalues, we have the desired conclusion when $\dim V = 2$.

Suppose now that $\dim V > 2$ and the desired result holds for all real vector spaces with smaller dimension. If $T$ has an eigenvalue, let $U$ be a one-dimensional subspace of $V$ that is invariant under $T$; otherwise let $U$ be a two-dimensional subspace of $V$ that is invariant under $T$ (5.24 guarantees that we can choose $U$ in this fashion). Choose any basis of $U$ and let $A_1$ denote the matrix of $T|_U$ with respect to this basis. If $A_1$ is a 2-by-2 matrix, then $T$ has no eigenvalues (otherwise we would have chosen $U$ to be one-dimensional) and thus $T|_U$ has no eigenvalues. Hence if $A_1$ is a 2-by-2 matrix, then $A_1$ has no eigenvalues (see 9.1).

Let $W$ be any subspace of $V$ such that

$$V = U \oplus W;$$

2.13 guarantees that such a $W$ exists. Because $W$ has dimension less than the dimension of $V$, we would like to apply our induction hypothesis to $T|_W$. However, $W$ might not be invariant under $T$, meaning that $T|_W$ might not be an operator on $W$. We will compose with the projection $P_{W,U}$ to get an operator on $W$. Specifically, define $S \in \mathcal{L}(W)$ by

*Recall that if $v = w + u$, where $w \in W$ and $u \in U$, then $P_{W,U}v = w$.*

$$Sw = P_{W,U}(Tw)$$

for $w \in W$. Note that

**9.6**
$$\begin{aligned} Tw &= P_{U,W}(Tw) + P_{W,U}(Tw) \\ &= P_{U,W}(Tw) + Sw \end{aligned}$$

for every $w \in W$.

By our induction hypothesis, there is a basis of $W$ with respect to which $S$ has a block upper-triangular matrix of the form

$$\begin{bmatrix} A_2 & & * \\ & \ddots & \\ 0 & & A_m \end{bmatrix},$$

where each $A_j$ is a 1-by-1 matrix or a 2-by-2 matrix with no eigenvalues. Adjoin this basis of $W$ to the basis of $U$ chosen above, getting a basis of $V$. A minute's thought should convince you (use 9.6) that the matrix of $T$ with respect to this basis is a block upper-triangular matrix of the form 9.5, completing the proof.  ∎

# *The Characteristic Polynomial*

For operators on complex vector spaces, we defined characteristic polynomials and developed their properties by making use of upper-triangular matrices. In this section we will carry out a similar procedure for operators on real vector spaces. Instead of upper-triangular matrices, we will have to use the block upper-triangular matrices furnished by the last theorem.

In the last chapter, we did not define the characteristic polynomial of a square matrix with complex entries because our emphasis is on operators rather than on matrices. However, to understand operators on real vector spaces, we will need to define characteristic polynomials of 1-by-1 and 2-by-2 matrices with real entries. Then, using block-upper triangular matrices with blocks of size at most 2-by-2 on the diagonal, we will be able to define the characteristic polynomial of an operator on a real vector space.

To motivate the definition of characteristic polynomials of square matrices, we would like the following to be true (think about the Cayley-Hamilton theorem; see 8.20): if $T \in \mathcal{L}(V)$ has matrix $A$ with respect to some basis of $V$ and $q$ is the characteristic polynomial of $A$, then $q(T) = 0$.

Let's begin with the trivial case of 1-by-1 matrices. Suppose $V$ is a real vector space with dimension 1 and $T \in \mathcal{L}(V)$. If $[\lambda]$ equals the matrix of $T$ with respect to some basis of $V$, then $T$ equals $\lambda I$. Thus if we let $q$ be the degree 1 polynomial defined by $q(x) = x - \lambda$, then $q(T) = 0$. Hence we define the characteristic polynomial of $[\lambda]$ to be $x - \lambda$.

Now let's look at 2-by-2 matrices with real entries. Suppose $V$ is a real vector space with dimension 2 and $T \in \mathcal{L}(V)$. Suppose

$$\begin{bmatrix} a & c \\ b & d \end{bmatrix}$$

is the matrix of $T$ with respect to some basis $(v_1, v_2)$ of $V$. We seek a monic polynomial $q$ of degree 2 such that $q(T) = 0$. If $b = 0$, then the matrix above is upper triangular. If in addition we were dealing with a complex vector space, then we would know that $T$ has characteristic polynomial $(z - a)(z - d)$. Thus a reasonable candidate might be $(x - a)(x - d)$, where we use $x$ instead of $z$ to emphasize that now we are working on a real vector space. Let's see if the polynomial

$(x - a)(x - d)$, when applied to $T$, gives 0 even when $b \neq 0$. We have

$$(T - aI)(T - dI)v_1 = (T - dI)(T - aI)v_1 = (T - dI)(bv_2) = bcv_1$$

and

$$(T - aI)(T - dI)v_2 = (T - aI)(cv_1) = bcv_2.$$

Thus $(T - aI)(T - dI)$ is not equal to 0 unless $bc = 0$. However, the equations above show that $(T - aI)(T - dI) - bcI = 0$ (because this operator equals 0 on a basis, it must equal 0 on $V$). Thus if $q(x) = (x - a)(x - d) - bc$, then $q(T) = 0$.

Motivated by the previous paragraph, we define the ***characteristic polynomial*** of a 2-by-2 matrix $\left[\begin{smallmatrix} a & c \\ b & d \end{smallmatrix}\right]$ to be $(x - a)(x - d) - bc$. Here we are concerned only with matrices with real entries. The next result shows that we have found the only reasonable definition for the characteristic polynomial of a 2-by-2 matrix.

**9.7**     **Proposition:** *Suppose $V$ is a real vector space with dimension 2 and $T \in \mathcal{L}(V)$ has no eigenvalues. Let $p \in \mathcal{P}(\mathbf{R})$ be a monic polynomial with degree 2. Suppose $A$ is the matrix of $T$ with respect to some basis of $V$.*

(a)     *If $p$ equals the characteristic polynomial of $A$, then $p(T) = 0$.*

(b)     *If $p$ does not equal the characteristic polynomial of $A$, then $p(T)$ is invertible.*

PROOF: We already proved (a) in our discussion above. To prove (b), let $q$ denote the characteristic polynomial of $A$ and suppose that $p \neq q$. We can write $p(x) = x^2 + \alpha_1 x + \beta_1$ and $q(x) = x^2 + \alpha_2 x + \beta_2$ for some $\alpha_1, \beta_1, \alpha_2, \beta_2 \in \mathbf{R}$. Now

$$p(T) = p(T) - q(T) = (\alpha_1 - \alpha_2)T + (\beta_1 - \beta_2)I.$$

If $\alpha_1 = \alpha_2$, then $\beta_1 \neq \beta_2$ (otherwise we would have $p = q$). Thus if $\alpha_1 = \alpha_2$, then $p(T)$ is a nonzero multiple of the identity and hence is invertible, as desired. If $\alpha_1 \neq \alpha_2$, then

$$p(T) = (\alpha_1 - \alpha_2)\left(T - \frac{\beta_2 - \beta_1}{\alpha_1 - \alpha_2}I\right),$$

which is an invertible operator because $T$ has no eigenvalues. Thus (b) holds.    ∎

*Part (b) of this proposition would be false without the hypothesis that $T$ has no eigenvalues. For example, define $T \in \mathcal{L}(\mathbf{R}^2)$ by $T(x_1, x_2) = (0, x_2)$. Take $p(x) = x(x - 2)$. Then $p$ is not the characteristic polynomial of the matrix of $T$ with respect to the standard basis, but $p(T)$ is not invertible.*

Suppose $V$ is a real vector space with dimension 2 and $T \in \mathcal{L}(V)$ has no eigenvalues. The last proposition shows that there is precisely one monic polynomial with degree 2 that when applied to $T$ gives 0. Thus, though $T$ may have different matrices with respect to different bases, each of these matrices must have the same characteristic polynomial. For example, consider $T \in \mathcal{L}(\mathbf{R}^2)$ defined by

**9.8** $$T(x_1, x_2) = (3x_1 + 5x_2, -2x_1 - x_2).$$

The matrix of $T$ with respect to the standard basis of $\mathbf{R}^2$ is

$$\begin{bmatrix} 3 & 5 \\ -2 & -1 \end{bmatrix}.$$

The characteristic polynomial of this matrix is $(x - 3)(x + 1) + 2 \cdot 5$, which equals $x^2 - 2x + 7$. As you should verify, the matrix of $T$ with respect to the basis $((-2, 1), (1, 2))$ equals

$$\begin{bmatrix} 1 & -6 \\ 1 & 1 \end{bmatrix}.$$

The characteristic polynomial of this matrix is $(x - 1)(x - 1) + 1 \cdot 6$, which equals $x^2 - 2x + 7$, the same result we obtained by using the standard basis.

When analyzing upper-triangular matrices of an operator $T$ on a complex vector space $V$, we found that subspaces of the form

$$\text{null}(T - \lambda I)^{\dim V}$$

played a key role (see 8.10). Those spaces will also play a role in studying operators on real vector spaces, but because we must now consider block upper-triangular matrices with 2-by-2 blocks, subspaces of the form

$$\text{null}(T^2 + \alpha T + \beta I)^{\dim V}$$

will also play a key role. To get started, let's look at one- and two-dimensional real vector spaces.

First suppose that $V$ is a one-dimensional real vector space and that $T \in \mathcal{L}(V)$. If $\lambda \in \mathbf{R}$, then $\text{null}(T - \lambda I)$ equals $V$ if $\lambda$ is an eigenvalue of $T$ and $\{0\}$ otherwise. If $\alpha, \beta \in \mathbf{R}$ with $\alpha^2 < 4\beta$, then

$$\text{null}(T^2 + \alpha T + \beta I) = \{0\}.$$

(Proof: Because $V$ is one-dimensional, there is a constant $\lambda \in \mathbf{R}$ such that $Tv = \lambda v$ for all $v \in V$. Thus $(T^2 + \alpha T + \beta I)v = (\lambda^2 + \alpha\lambda + \beta)v$. However, the inequality $\alpha^2 < 4\beta$ implies that $\lambda^2 + \alpha\lambda + \beta \neq 0$, and thus $\text{null}(T^2 + \alpha T + \beta I) = \{0\}$.)

*Recall that $\alpha^2 < 4\beta$ implies that $x^2 + \alpha x + \beta$ has no real roots; see 4.11.*

Now suppose $V$ is a two-dimensional real vector space and $T \in \mathcal{L}(V)$ has no eigenvalues. If $\lambda \in \mathbf{R}$, then $\text{null}(T - \lambda I)$ equals $\{0\}$ (because $T$ has no eigenvalues). If $\alpha, \beta \in \mathbf{R}$ with $\alpha^2 < 4\beta$, then $\text{null}(T^2 + \alpha T + \beta I)$ equals $V$ if $x^2 + \alpha x + \beta$ is the characteristic polynomial of the matrix of $T$ with respect to some (or equivalently, every) basis of $V$ and equals $\{0\}$ otherwise (by 9.7). Note that for this operator, there is no middle ground—the null space of $T^2 + \alpha T + \beta I$ is either $\{0\}$ or the whole space; it cannot be one-dimensional.

Now suppose that $V$ is a real vector space of any dimension and $T \in \mathcal{L}(V)$. We know that $V$ has a basis with respect to which $T$ has a block upper-triangular matrix with blocks on the diagonal of size at most 2-by-2 (see 9.4). In general, this matrix is not unique—$V$ may have many different bases with respect to which $T$ has a block upper-triangular matrix of this form, and with respect to these different bases we may get different block upper-triangular matrices.

We encountered a similar situation when dealing with complex vector spaces and upper-triangular matrices. In that case, though we might get different upper-triangular matrices with respect to the different bases, the entries on the diagonal were always the same (though possibly in a different order). Might a similar property hold for real vector spaces and block upper-triangular matrices? Specifically, is the number of times a given 2-by-2 matrix appears on the diagonal of a block upper-triangular matrix of $T$ independent of which basis is chosen? Unfortunately this question has a negative answer. For example, the operator $T \in \mathcal{L}(\mathbf{R}^2)$ defined by 9.8 has two different 2-by-2 matrices, as we saw above.

Though the number of times a particular 2-by-2 matrix might appear on the diagonal of a block upper-triangular matrix of $T$ can depend on the choice of basis, if we look at characteristic polynomials instead of the actual matrices, we find that the number of times a particular characteristic polynomial appears is independent of the choice of basis. This is the content of the following theorem, which will be our key tool in analyzing the structure of an operator on a real vector space.

**9.9**     **Theorem:**  *Suppose V is a real vector space and* $T \in \mathcal{L}(V)$. *Suppose that with respect to some basis of V, the matrix of T is*

**9.10**
$$\begin{bmatrix} A_1 & & * \\ & \ddots & \\ 0 & & A_m \end{bmatrix},$$

*where each* $A_j$ *is a 1-by-1 matrix or a 2-by-2 matrix with no eigenvalues.*

*This result implies that* $\text{null}(T^2 + \alpha T + \beta I)^{\dim V}$ *must have even dimension.*

(a)     *If* $\lambda \in \mathbf{R}$, *then precisely* $\dim \text{null}(T - \lambda I)^{\dim V}$ *of the matrices* $A_1, \ldots, A_m$ *equal the 1-by-1 matrix* $[\lambda]$.

(b)     *If* $\alpha, \beta \in \mathbf{R}$ *satisfy* $\alpha^2 < 4\beta$, *then precisely*

$$\frac{\dim \text{null}(T^2 + \alpha T + \beta I)^{\dim V}}{2}$$

*of the matrices* $A_1, \ldots, A_m$ *have characteristic polynomial equal to* $x^2 + \alpha x + \beta$.

*This proof uses the same ideas as the proof of the analogous result on complex vector spaces (8.10). As usual, the real case is slightly more complicated but requires no new creativity.*

PROOF:  We will construct one proof that can be used to prove both (a) and (b). To do this, let $\lambda, \alpha, \beta \in \mathbf{R}$ with $\alpha^2 < 4\beta$. Define $p \in \mathcal{P}(\mathbf{R})$ by

$$p(x) = \begin{cases} x - \lambda & \text{if we are trying to prove (a);} \\ x^2 + \alpha x + \beta & \text{if we are trying to prove (b).} \end{cases}$$

Let $d$ denote the degree of $p$. Thus $d = 1$ if we are trying to prove (a) and $d = 2$ if we are trying to prove (b).

We will prove this theorem by induction on $m$, the number of blocks along the diagonal of 9.10. If $m = 1$, then $\dim V = 1$ or $\dim V = 2$; the discussion preceding this theorem then implies that the desired result holds. Thus we can assume that $m > 1$ and that the desired result holds when $m$ is replaced with $m - 1$.

For convenience let $n = \dim V$. Consider a basis of $V$ with respect to which $T$ has the block upper-triangular matrix 9.10. Let $U_j$ denote the span of the basis vectors corresponding to $A_j$. Thus $\dim U_j = 1$ if $A_j$ is a 1-by-1 matrix and $\dim U_j = 2$ if $A_j$ is a 2-by-2 matrix. Let $U = U_1 + \cdots + U_{m-1}$. Clearly $U$ is invariant under $T$ and the matrix of $T|_U$ with respect to the obvious basis (obtained from the basis vectors corresponding to $A_1, \ldots, A_{m-1}$) is

**9.11**
$$\begin{bmatrix} A_1 & & * \\ & \ddots & \\ 0 & & A_{m-1} \end{bmatrix}.$$

Thus, by our induction hypothesis,

**9.12**      precisely $(1/d)$ dim null $p(T|_U)^n$ of the matrices
             $A_1, \ldots, A_{m-1}$ have characteristic polynomial $p$.

Actually the induction hypothesis gives 9.12 with exponent $\dim U$ instead of $n$, but then we can replace $\dim U$ with $n$ (by 8.6) to get the statement above.

Suppose $u_m \in U_m$. Let $S \in \mathcal{L}(U_m)$ be the operator whose matrix (with respect to the basis corresponding to $U_m$) equals $A_m$. In particular, $Su_m = P_{U_m,U} Tu_m$. Now

$$Tu_m = P_{U,U_m} Tu_m + P_{U_m,U} Tu_m$$
$$= *_U + Su_m,$$

where $*_U$ denotes a vector in $U$. Note that $Su_m \in U_m$; thus applying $T$ to both sides of the equation above gives

$$T^2 u_m = *_U + S^2 u_m,$$

where again $*_U$ denotes a vector in $U$, though perhaps a different vector than the previous usage of $*_U$ (the notation $*_U$ is used when we want to emphasize that we have a vector in $U$ but we do not care which particular vector—each time the notation $*_U$ is used, it may denote a different vector in $U$). The last two equations show that

**9.13**                 $p(T)u_m = *_U + p(S)u_m$

for some $*_U \in U$. Note that $p(S)u_m \in U_m$; thus iterating the last equation gives

**9.14**                 $p(T)^n u_m = *_U + p(S)^n u_m$

for some $*_U \in U$.

The proof now breaks into two cases. First consider the case where the characteristic polynomial of $A_m$ does not equal $p$. We will show that in this case

**9.15**                 null $p(T)^n \subset U$.

Once this has been verified, we will know that

$$\text{null } p(T)^n = \text{null } p(T|_U)^n,$$

and hence 9.12 will tell us that precisely $(1/d) \dim \text{null} \, p(T)^n$ of the matrices $A_1, \ldots, A_m$ have characteristic polynomial $p$, completing the proof in the case where the characteristic polynomial of $A_m$ does not equal $p$.

To prove 9.15 (still assuming that the characteristic polynomial of $A_m$ does not equal $p$), suppose $v \in \text{null} \, p(T)^n$. We can write $v$ in the form $v = u + u_m$, where $u \in U$ and $u_m \in U_m$. Using 9.14, we have

$$0 = p(T)^n v = p(T)^n u + p(T)^n u_m = p(T)^n u + *_U + p(S)^n u_m$$

for some $*_U \in U$. Because the vectors $p(T)^n u$ and $*_U$ are in $U$ and $p(S)^n u_m \in U_m$, this implies that $p(S)^n u_m = 0$. However, $p(S)$ is invertible (see the discussion preceding this theorem about one- and two-dimensional subspaces and note that $\dim U_m \leq 2$), so $u_m = 0$. Thus $v = u \in U$, completing the proof of 9.15.

Now consider the case where the characteristic polynomial of $A_m$ equals $p$. Note that this implies $\dim U_m = d$. We will show that

**9.16**              $\dim \text{null} \, p(T)^n = \dim \text{null} \, p(T|_U)^n + d,$

which along with 9.12 will complete the proof.

Using the formula for the dimension of the sum of two subspaces (2.18), we have

$$\dim \text{null} \, p(T)^n = \dim(U \cap \text{null} \, p(T)^n) + \dim(U + \text{null} \, p(T)^n) - \dim U$$
$$= \dim \text{null} \, p(T|_U)^n + \dim(U + \text{null} \, p(T)^n) - (n - d).$$

If $U + \text{null} \, p(T)^n = V$, then $\dim(U + \text{null} \, p(T)^n) = n$, which when combined with the last formula above for $\dim \text{null} \, p(T)^n$ would give 9.16, as desired. Thus we will finish by showing that $U + \text{null} \, p(T)^n = V$.

To prove that $U + \text{null} \, p(T)^n = V$, suppose $u_m \in U_m$. Because the characteristic polynomial of the matrix of $S$ (namely, $A_m$) equals $p$, we have $p(S) = 0$. Thus $p(T)u_m \in U$ (from 9.13). Now

$$p(T)^n u_m = p(T)^{n-1}(p(T)u_m) \in \text{range} \, p(T|_U)^{n-1} = \text{range} \, p(T|_U)^n,$$

where the last equality comes from 8.9. Thus we can choose $u \in U$ such that $p(T)^n u_m = p(T|_U)^n u$. Now

$$p(T)^n(u_m - u) = p(T)^n u_m - p(T)^n u$$
$$= p(T)^n u_m - p(T|_U)^n u$$
$$= 0.$$

Thus $u_m - u \in \text{null} \, p(T)^n$, and hence $u_m$, which equals $u + (u_m - u)$, is in $U + \text{null} \, p(T)^n$. In other words, $U_m \subset U + \text{null} \, p(T)^n$. Therefore $V = U + U_m \subset U + \text{null} \, p(T)^n$, and hence $U + \text{null} \, p(T)^n = V$, completing the proof. ∎

As we saw in the last chapter, the eigenvalues of an operator on a complex vector space provide the key to analyzing the structure of the operator. On a real vector space, an operator may have fewer eigenvalues, counting multiplicity, than the dimension of the vector space. The previous theorem suggests a definition that makes up for this deficiency. We will see that the definition given in the next paragraph helps make operator theory on real vector spaces resemble operator theory on complex vector spaces.

Suppose $V$ is a real vector space and $T \in \mathcal{L}(V)$. An ordered pair $(\alpha, \beta)$ of real numbers is called an ***eigenpair*** of $T$ if $\alpha^2 < 4\beta$ and

$$T^2 + \alpha T + \beta I$$

*Though the word **eigenpair** was chosen to be consistent with the word **eigenvalue**, this terminology is not in widespread use.*

is not injective. The previous theorem shows that $T$ can have only finitely many eigenpairs because each eigenpair corresponds to the characteristic polynomial of a 2-by-2 matrix on the diagonal of 9.10 and there is room for only finitely many such matrices along that diagonal. Guided by 9.9, we define the ***multiplicity*** of an eigenpair $(\alpha, \beta)$ of $T$ to be

$$\frac{\dim \text{null}(T^2 + \alpha T + \beta I)^{\dim V}}{2}.$$

From 9.9, we see that the multiplicity of $(\alpha, \beta)$ equals the number of times that $x^2 + \alpha x + \beta$ is the characteristic polynomial of a 2-by-2 matrix on the diagonal of 9.10.

As an example, consider the operator $T \in \mathcal{L}(\mathbf{R}^3)$ whose matrix (with respect to the standard basis) equals

$$\begin{bmatrix} 3 & -1 & -2 \\ 3 & 2 & -3 \\ 1 & 2 & 0 \end{bmatrix}.$$

You should verify that $(-4, 13)$ is an eigenpair of $T$ with multiplicity 1; note that $T^2 - 4T + 13I$ is not injective because $(-1, 0, 1)$ and $(1, 1, 0)$ are in its null space. Without doing any calculations, you should verify that $T$ has no other eigenpairs (use 9.9). You should also verify that 1 is an eigenvalue of $T$ with multiplicity 1, with corresponding eigenvector $(1, 0, 1)$, and that $T$ has no other eigenvalues.

In the example above, the sum of the multiplicities of the eigenvalues of $T$ plus twice the multiplicities of the eigenpairs of $T$ equals 3, which is the dimension of the domain of $T$. The next proposition shows that this always happens on a real vector space.

*This proposition shows that though an operator on a real vector space may have no eigenvalues, or it may have no eigenpairs, it cannot be lacking in both these useful objects. It also shows that an operator on a real vector space $V$ can have at most $(\dim V)/2$ distinct eigenpairs.*

**9.17   Proposition:**   *If $V$ is a real vector space and $T \in \mathcal{L}(V)$, then the sum of the multiplicities of all the eigenvalues of $T$ plus the sum of twice the multiplicities of all the eigenpairs of $T$ equals $\dim V$.*

PROOF:   Suppose $V$ is a real vector space and $T \in \mathcal{L}(V)$. Then there is a basis of $V$ with respect to which the matrix of $T$ is as in 9.9. The multiplicity of an eigenvalue $\lambda$ equals the number of times the 1-by-1 matrix $[\lambda]$ appears on the diagonal of this matrix (from 9.9). The multiplicity of an eigenpair $(\alpha, \beta)$ equals the number of times $x^2 + \alpha x + \beta$ is the characteristic polynomial of a 2-by-2 matrix on the diagonal of this matrix (from 9.9). Because the diagonal of this matrix has length $\dim V$, the sum of the multiplicities of all the eigenvalues of $T$ plus the sum of twice the multiplicities of all the eigenpairs of $T$ must equal $\dim V$.   ∎

Suppose $V$ is a real vector space and $T \in \mathcal{L}(V)$. With respect to some basis of $V$, $T$ has a block upper-triangular matrix of the form

**9.18**
$$\begin{bmatrix} A_1 & & * \\ & \ddots & \\ 0 & & A_m \end{bmatrix},$$

where each $A_j$ is a 1-by-1 matrix or a 2-by-2 matrix with no eigenvalues (see 9.4). We define the ***characteristic polynomial*** of $T$ to be the product of the characteristic polynomials of $A_1, \ldots, A_m$. Explicitly, for each $j$, define $q_j \in \mathcal{P}(\mathbf{R})$ by

**9.19**      $q_j(x) = \begin{cases} x - \lambda & \text{if } A_j \text{ equals } [\lambda]; \\ (x-a)(x-d) - bc & \text{if } A_j \text{ equals } [\begin{smallmatrix} a & c \\ b & d \end{smallmatrix}]. \end{cases}$

*Note that the roots of the characteristic polynomial of $T$ equal the eigenvalues of $T$, as was true on complex vector spaces.*

Then the characteristic polynomial of $T$ is

$$q_1(x) \ldots q_m(x).$$

Clearly the characteristic polynomial of $T$ has degree $\dim V$. Furthermore, 9.9 insures that the characteristic polynomial of $T$ depends only on $T$ and not on the choice of a particular basis.

Now we can prove a result that was promised in the last chapter, where we proved the analogous theorem (8.20) for operators on complex vector spaces.

**9.20  Cayley-Hamilton Theorem:** *Suppose V is a real vector space and $T \in \mathcal{L}(V)$. Let $q$ denote the characteristic polynomial of $T$. Then $q(T) = 0$.*

PROOF:  Choose a basis of $V$ with respect to which $T$ has a block upper-triangular matrix of the form 9.18, where each $A_j$ is a 1-by-1 matrix or a 2-by-2 matrix with no eigenvalues. Suppose $U_j$ is the one- or two-dimensional subspace spanned by the basis vectors corresponding to $A_j$. Define $q_j$ as in 9.19. To prove that $q(T) = 0$, we need only show that $q(T)|_{U_j} = 0$ for $j = 1, \ldots, m$. To do this, it suffices to show that

*This proof uses the same ideas as the proof of the analogous result on complex vector spaces (8.20).*

**9.21**
$$q_1(T) \ldots q_j(T)|_{U_j} = 0$$

for $j = 1, \ldots, m$.

We will prove 9.21 by induction on $j$. To get started, suppose that $j = 1$. Because $\mathcal{M}(T)$ is given by 9.18, we have $q_1(T)|_{U_1} = 0$ (obvious if $\dim U_1 = 1$; from 9.7(a) if $\dim U_1 = 2$), giving 9.21 when $j = 1$.

Now suppose that $1 < j \le n$ and that

$$0 = q_1(T)|_{U_1}$$
$$0 = q_1(T)q_2(T)|_{U_2}$$
$$\vdots$$
$$0 = q_1(T) \ldots q_{j-1}(T)|_{U_{j-1}}.$$

If $v \in U_j$, then from 9.18 we see that

$$q_j(T)v = u + q_j(S)v,$$

where $u \in U_1 + \cdots + U_{j-1}$ and $S \in \mathcal{L}(U_j)$ has characteristic polynomial $q_j$. Because $q_j(S) = 0$ (obvious if $\dim U_j = 1$; from 9.7(a) if $\dim U_j = 2$), the equation above shows that

$$q_j(T)v \in U_1 + \cdots + U_{j-1}$$

whenever $v \in U_j$. Thus, by our induction hypothesis, $q_1(T) \ldots q_{j-1}(T)$ applied to $q_j(T)v$ gives 0 whenever $v \in U_j$. In other words, 9.21 holds, completing the proof.  ∎

Suppose $V$ is a real vector space and $T \in \mathcal{L}(V)$. Clearly the Cayley-Hamilton theorem (9.20) implies that the minimal polynomial of $T$ has degree at most $\dim V$, as was the case on complex vector spaces. If the degree of the minimal polynomial of $T$ equals $\dim V$, then, as was also the case on complex vector spaces, the minimal polynomial of $T$ must equal the characteristic polynomial of $T$. This follows from the Cayley-Hamilton theorem (9.20) and 8.34.

Finally, we can now prove a major structure theorem about operators on real vector spaces. The theorem below should be compared to 8.23, the corresponding result on complex vector spaces.

*Either $m$ or $M$ might be 0.*

**9.22   Theorem:** *Suppose $V$ is a real vector space and $T \in \mathcal{L}(V)$. Let $\lambda_1, \ldots, \lambda_m$ be the distinct eigenvalues of $T$, with $U_1, \ldots, U_m$ the corresponding sets of generalized eigenvectors. Let $(\alpha_1, \beta_1), \ldots, (\alpha_M, \beta_M)$ be the distinct eigenpairs of $T$ and let $V_j = \operatorname{null}(T^2 + \alpha_j T + \beta_j I)^{\dim V}$. Then*

(a)     $V = U_1 \oplus \cdots \oplus U_m \oplus V_1 \oplus \cdots \oplus V_M$;

(b)     *each $U_j$ and each $V_j$ is invariant under $T$;*

(c)     *each $(T - \lambda_j I)|_{U_j}$ and each $(T^2 + \alpha_j T + \beta_j I)|_{V_j}$ is nilpotent.*

*This proof uses the same ideas as the proof of the analogous result on complex vector spaces (8.23).*

PROOF: From 8.22, we get (b). Clearly (c) follows from the definitions.

To prove (a), recall that $\dim U_j$ equals the multiplicity of $\lambda_j$ as an eigenvalue of $T$ and $\dim V_j$ equals twice the multiplicity of $(\alpha_j, \beta_j)$ as an eigenpair of $T$. Thus

**9.23**        $\dim V = \dim U_1 + \cdots + \dim U_m + \dim V_1 + \cdots + V_M$;

this follows from 9.17. Let $U = U_1 + \cdots + U_m + V_1 + \cdots + V_M$. Note that $U$ is invariant under $T$. Thus we can define $S \in \mathcal{L}(U)$ by

$$S = T|_U.$$

Note that $S$ has the same eigenvalues, with the same multiplicities, as $T$ because all the generalized eigenvectors of $T$ are in $U$, the domain of $S$. Similarly, $S$ has the same eigenpairs, with the same multiplicities, as $T$. Thus applying 9.17 to $S$, we get

$$\dim U = \dim U_1 + \cdots + \dim U_m + \dim V_1 + \cdots + V_M.$$

This equation, along with 9.23, shows that $\dim V = \dim U$. Because $U$ is a subspace of $V$, this implies that $V = U$. In other words,

$$V = U_1 + \cdots + U_m + V_1 + \cdots + V_M.$$

This equation, along with 9.23, allows us to use 2.19 to conclude that (a) holds, completing the proof.                                      ∎

# *Exercises*

1.  Prove that 1 is an eigenvalue of every square matrix with the property that the sum of the entries in each row equals 1.

2.  Consider a 2-by-2 matrix of real numbers

    $$A = \begin{bmatrix} a & c \\ b & d \end{bmatrix}.$$

    Prove that $A$ has an eigenvalue (in $\mathbf{R}$) if and only if

    $$(a - d)^2 + 4bc \geq 0.$$

3.  Suppose $A$ is a block diagonal matrix

    $$A = \begin{bmatrix} A_1 & & 0 \\ & \ddots & \\ 0 & & A_m \end{bmatrix},$$

    where each $A_j$ is a square matrix. Prove that the set of eigenvalues of $A$ equals the union of the eigenvalues of $A_1, \dots, A_m$.

*Clearly Exercise 4 is a stronger statement than Exercise 3. Even so, you may want to do Exercise 3 first because it is easier than Exercise 4.*

4.  Suppose $A$ is a block upper-triangular matrix

    $$A = \begin{bmatrix} A_1 & & * \\ & \ddots & \\ 0 & & A_m \end{bmatrix},$$

    where each $A_j$ is a square matrix. Prove that the set of eigenvalues of $A$ equals the union of the eigenvalues of $A_1, \dots, A_m$.

5.  Suppose $V$ is a real vector space and $T \in \mathcal{L}(V)$. Suppose $\alpha, \beta \in \mathbf{R}$ are such that $T^2 + \alpha T + \beta I = 0$. Prove that $T$ has an eigenvalue if and only if $\alpha^2 \geq 4\beta$.

6.  Suppose $V$ is a real inner-product space and $T \in \mathcal{L}(V)$. Prove that there is an orthonormal basis of $V$ with respect to which $T$ has a block upper-triangular matrix

    $$\begin{bmatrix} A_1 & & * \\ & \ddots & \\ 0 & & A_m \end{bmatrix},$$

    where each $A_j$ is a 1-by-1 matrix or a 2-by-2 matrix with no eigenvalues.

7. Prove that if $T \in \mathcal{L}(V)$ and $j$ is a positive integer such that $j \le \dim V$, then $T$ has an invariant subspace whose dimension equals $j - 1$ or $j$.

8. Prove that there does not exist an operator $T \in \mathcal{L}(\mathbf{R}^7)$ such that $T^2 + T + I$ is nilpotent.

9. Give an example of an operator $T \in \mathcal{L}(\mathbf{C}^7)$ such that $T^2 + T + I$ is nilpotent.

10. Suppose $V$ is a real vector space and $T \in \mathcal{L}(V)$. Suppose $\alpha, \beta \in \mathbf{R}$ are such that $\alpha^2 < 4\beta$. Prove that

$$\text{null}(T^2 + \alpha T + \beta I)^k$$

has even dimension for every positive integer $k$.

11. Suppose $V$ is a real vector space and $T \in \mathcal{L}(V)$. Suppose $\alpha, \beta \in \mathbf{R}$ are such that $\alpha^2 < 4\beta$ and $T^2 + \alpha T + \beta I$ is nilpotent. Prove that $\dim V$ is even and

$$(T^2 + \alpha T + \beta I)^{\dim V/2} = 0.$$

12. Prove that if $T \in \mathcal{L}(\mathbf{R}^3)$ and $5, 7$ are eigenvalues of $T$, then $T$ has no eigenpairs.

13. Suppose $V$ is a real vector space with $\dim V = n$ and $T \in \mathcal{L}(V)$ is such that
$$\text{null } T^{n-2} \ne \text{null } T^{n-1}.$$

Prove that $T$ has at most two distinct eigenvalues and that $T$ has no eigenpairs.

14. Suppose $V$ is a vector space with dimension 2 and $T \in \mathcal{L}(V)$. Prove that if

$$\begin{bmatrix} a & c \\ b & d \end{bmatrix}$$

is the matrix of $T$ with respect to some basis of $V$, then the characteristic polynomial of $T$ equals $(z - a)(z - d) - bc$.

*You do not need to find the eigenvalues of T to do this exercise. As usual unless otherwise specified, here V may be a real or complex vector space.*

15. Suppose $V$ is a real inner-product space and $S \in \mathcal{L}(V)$ is an isometry. Prove that if $(\alpha, \beta)$ is an eigenpair of $S$, then $\beta = 1$.

# CHAPTER 10

# *Trace and Determinant*

Throughout this book our emphasis has been on linear maps and operators rather than on matrices. In this chapter we pay more attention to matrices as we define and discuss traces and determinants. Determinants appear only at the end of this book because we replaced their usual applications in linear algebra (the definition of the characteristic polynomial and the proof that operators on complex vector spaces have eigenvalues) with more natural techniques. The book concludes with an explanation of the important role played by determinants in the theory of volume and integration.

> Recall that **F** denotes **R** or **C**.
> Also, $V$ is a finite-dimensional, nonzero vector space over **F**.

# *Change of Basis*

The matrix of an operator $T \in \mathcal{L}(V)$ depends on a choice of basis of $V$. Two different bases of $V$ may give different matrices of $T$. In this section we will learn how these matrices are related. This information will help us find formulas for the trace and determinant of $T$ later in this chapter.

With respect to any basis of $V$, the identity operator $I \in \mathcal{L}(V)$ has a diagonal matrix

$$\begin{bmatrix} 1 & & 0 \\ & \ddots & \\ 0 & & 1 \end{bmatrix}.$$

This matrix is called the ***identity matrix*** and is denoted $I$. Note that we use the symbol $I$ to denote the identity operator (on all vector spaces) and the identity matrix (of all possible sizes). You should always be able to tell from the context which particular meaning of $I$ is intended. For example, consider the equation

$$\mathcal{M}(I) = I;$$

on the left side $I$ denotes the identity operator and on the right side $I$ denotes the identity matrix.

If $A$ is a square matrix (with entries in $\mathbf{F}$, as usual) with the same size as $I$, then $AI = IA = A$, as you should verify. A square matrix $A$ is called ***invertible*** if there is a square matrix $B$ of the same size such that $AB = BA = I$, and we call $B$ an ***inverse*** of $A$. To prove that $A$ has at most one inverse, suppose $B$ and $B'$ are inverses of $A$. Then

$$B = BI = B(AB') = (BA)B' = IB' = B',$$

*Some mathematicians use the terms **nonsingular**, which means the same as invertible, and **singular**, which means the same as noninvertible.*

and hence $B = B'$, as desired. Because an inverse is unique, we can use the notation $A^{-1}$ to denote the inverse of $A$ (if $A$ is invertible). In other words, if $A$ is invertible, then $A^{-1}$ is the unique matrix of the same size such that $AA^{-1} = A^{-1}A = I$.

Recall that when discussing linear maps from one vector space to another in Chapter 3, we defined the matrix of a linear map with respect to two bases—one basis for the first vector space and another basis for the second vector space. When we study operators, which are linear maps from a vector space to itself, we almost always use the same basis for both vector spaces (after all, the two vector spaces in question are equal). Thus we usually refer to the matrix of an operator with respect to a basis, meaning that we are using one basis in two capacities. The next proposition is one of the rare cases where we need to use two different bases even though we have an operator from a vector space to itself.

Let's review how matrix multiplication interacts with multiplication of linear maps. Suppose that along with $V$ we have two other finite-dimensional vector spaces, say $U$ and $W$. Let $(u_1, \ldots, u_p)$ be a basis of $U$, let $(v_1, \ldots, v_n)$ be a basis of $V$, and let $(w_1, \ldots, w_m)$ be a basis of $W$. If $T \in \mathcal{L}(U, V)$ and $S \in \mathcal{L}(V, W)$, then $ST \in \mathcal{L}(U, W)$ and

**10.1** $\mathcal{M}(ST, (u_1, \ldots, u_p), (w_1, \ldots, w_m)) =$

$\qquad \mathcal{M}(S, (v_1, \ldots, v_n), (w_1, \ldots, w_m)) \mathcal{M}(T, (u_1, \ldots, u_m), (v_1, \ldots, v_n)).$

The equation above holds because we defined matrix multiplication to make it true—see 3.11 and the material following it.

The following proposition deals with the matrix of the identity operator when we use two different bases. Note that the $k^{\text{th}}$ column of $\mathcal{M}(I, (u_1, \ldots, u_n), (v_1, \ldots, v_n))$ consists of the scalars needed to write $u_k$ as a linear combination of the $v$'s. As an example of the proposition below, consider the bases $((4, 2), (5, 3))$ and $((1, 0), (0, 1))$ of $\mathbf{F}^2$. Obviously

$$\mathcal{M}\big(I, ((4, 2), (5, 3)), ((1, 0), (0, 1))\big) = \begin{bmatrix} 4 & 5 \\ 2 & 3 \end{bmatrix}.$$

The inverse of the matrix above is $\begin{bmatrix} 3/2 & -5/2 \\ -1 & 2 \end{bmatrix}$, as you should verify. Thus the proposition below implies that

$$\mathcal{M}\big(I, ((1, 0), (0, 1)), ((4, 2), (5, 3))\big) = \begin{bmatrix} 3/2 & -5/2 \\ -1 & 2 \end{bmatrix}.$$

**10.2  Proposition:** *If $(u_1, \ldots, u_n)$ and $(v_1, \ldots, v_n)$ are bases of $V$, then $\mathcal{M}(I, (u_1, \ldots, u_n), (v_1, \ldots, v_n))$ is invertible and*

$$\mathcal{M}(I, (u_1, \ldots, u_n), (v_1, \ldots, v_n))^{-1} = \mathcal{M}(I, (v_1, \ldots, v_n), (u_1, \ldots, u_n)).$$

PROOF:  In 10.1, replace $U$ and $W$ with $V$, replace $w_j$ with $u_j$, and replace $S$ and $T$ with $I$, getting

$$I = \mathcal{M}(I, (v_1, \ldots, v_n), (u_1, \ldots, u_n))\mathcal{M}(I, (u_1, \ldots, u_n), (v_1, \ldots, v_n)).$$

Now interchange the roles of the $u$'s and $v$'s, getting

$$I = \mathcal{M}(I, (u_1, \ldots, u_n), (v_1, \ldots, v_n))\mathcal{M}(I, (v_1, \ldots, v_n), (u_1, \ldots, u_n)).$$

These two equations give the desired result.                                    ∎

Now we can see how the matrix of $T$ changes when we change bases.

**10.3  Theorem:** *Suppose $T \in \mathcal{L}(V)$. Let $(u_1, \ldots, u_n)$ and $(v_1, \ldots, v_n)$ be bases of $V$. Let $A = \mathcal{M}(I, (u_1, \ldots, u_n), (v_1, \ldots, v_n))$. Then*

**10.4**        $\mathcal{M}(T, (u_1, \ldots, u_n)) = A^{-1}\mathcal{M}(T, (v_1, \ldots, v_n))A.$

PROOF:  In 10.1, replace $U$ and $W$ with $V$, replace $w_j$ with $v_j$, replace $T$ with $I$, and replace $S$ with $T$, getting

**10.5**        $\mathcal{M}(T, (u_1, \ldots, u_n), (v_1, \ldots, v_n)) = \mathcal{M}(T, (v_1, \ldots, v_n))A.$

Again use 10.1, this time replacing $U$ and $W$ with $V$, replacing $w_j$ with $u_j$, and replacing $S$ with $I$, getting

$$\mathcal{M}(T, (u_1, \ldots, u_n)) = A^{-1}\mathcal{M}(T, (u_1, \ldots, u_m), (v_1, \ldots, v_n)),$$

where we have used 10.2.  Substituting 10.5 into the equation above gives 10.4, completing the proof.                                    ∎

## *Trace*

Let's examine the characteristic polynomial more closely than we did in the last two chapters. If $V$ is an $n$-dimensional complex vector space and $T \in \mathcal{L}(V)$, then the characteristic polynomial of $T$ equals

$$(z - \lambda_1) \ldots (z - \lambda_n),$$

where $\lambda_1, \ldots, \lambda_n$ are the eigenvalues of $T$, repeated according to multiplicity. Expanding the polynomial above, we can write the characteristic polynomial of $T$ in the form

**10.6**        $z^n - (\lambda_1 + \cdots + \lambda_n)z^{n-1} + \cdots + (-1)^n(\lambda_1 \ldots \lambda_n).$

If $V$ is an $n$-dimensional real vector space and $T \in \mathcal{L}(V)$, then the characteristic polynomial of $T$ equals

$$(x - \lambda_1) \ldots (x - \lambda_m)(x^2 + \alpha_1 x + \beta_1) \ldots (x^2 + \alpha_M x + \beta_M),$$

*Here m or M might equal 0.*

where $\lambda_1, \ldots, \lambda_m$ are the eigenvalues of $T$ and $(\alpha_1, \beta_1), \ldots, (\alpha_M, \beta_M)$ are the eigenpairs of $T$, each repeated according to multiplicity. Expanding the polynomial above, we can write the characteristic polynomial of $T$ in the form

*Recall that a pair $(\alpha, \beta)$ of real numbers is an eigenpair of T if $\alpha^2 < 4\beta$ and*

**10.7**   $x^n - (\lambda_1 + \cdots + \lambda_m - \alpha_1 - \cdots - \alpha_m)x^{n-1} + \ldots$
$$+ (-1)^m (\lambda_1 \ldots \lambda_m \beta_1 \ldots \beta_M).$$

*$T^2 + \alpha T + \beta I$ is not injective.*

In this section we will study the coefficient of $z^{n-1}$ (usually denoted $x^{n-1}$ when we are dealing with a real vector space) in the characteristic polynomial. In the next section we will study the constant term in the characteristic polynomial.

For $T \in \mathcal{L}(V)$, the negative of the coefficient of $z^{n-1}$ (or $x^{n-1}$ for real vector spaces) in the characteristic polynomial of $T$ is called the **trace** of $T$, denoted trace $T$. If $V$ is a complex vector space, then 10.6 shows that trace $T$ equals the sum of the eigenvalues of $T$, counting multiplicity. If $V$ is a real vector space, then 10.7 shows that trace $T$ equals the sum of the eigenvalues of $T$ minus the sum of the first coordinates of the eigenpairs of $T$, each repeated according to multiplicity.

*Note that trace T depends only on T and not on a basis of V because the characteristic polynomial of T does not depend on a choice of basis.*

For example, suppose $T \in \mathcal{L}(\mathbf{C}^3)$ is the operator whose matrix is

**10.8**
$$\begin{bmatrix} 3 & -1 & -2 \\ 3 & 2 & -3 \\ 1 & 2 & 0 \end{bmatrix}.$$

Then the eigenvalues of $T$ are $1$, $2 + 3i$, and $2 - 3i$, each with multiplicity 1, as you can verify. Computing the sum of the eigenvalues, we have trace $T = 1 + (2 + 3i) + (2 - 3i)$; in other words, trace $T = 5$.

As another example, suppose $T \in \mathcal{L}(\mathbf{R}^3)$ is the operator whose matrix is also given by 10.8 (note that in the previous paragraph we were working on a complex vector space; now we are working on a real vector space). Then 1 is the only eigenvalue of $T$ (it has multiplicity 1) and $(-4, 13)$ is the only eigenpair of $T$ (it has multiplicity 1), as you should have verified in the last chapter (see page 205). Computing the sum of the eigenvalues minus the sum of the first coordinates of the eigenpairs, we have trace $T = 1 - (-4)$; in other words, trace $T = 5$.

The reason that the operators in the two previous examples have the same trace will become clear after we find a formula (valid on both complex and real vector spaces) for computing the trace of an operator from its matrix.

Most of the rest of this section is devoted to discovering how to calculate trace $T$ from the matrix of $T$ (with respect to an arbitrary basis). Let's start with the easiest situation. Suppose $V$ is a complex vector space, $T \in \mathcal{L}(V)$, and we choose a basis of $V$ with respect to which $T$ has an upper-triangular matrix $A$. Then the eigenvalues of $T$ are precisely the diagonal entries of $A$, repeated according to multiplicity (see 8.10). Thus trace $T$ equals the sum of the diagonal entries of $A$. The same formula works for the operator $T \in \mathcal{L}(\mathbf{F}^3)$ whose matrix is given by 10.8 and whose trace equals 5. Could such a simple formula be true in general?

We begin our investigation by considering $T \in \mathcal{L}(V)$ where $V$ is a real vector space. Choose a basis of $V$ with respect to which $T$ has a block upper-triangular matrix $\mathcal{M}(T)$, where each block on the diagonal is a 1-by-1 matrix containing an eigenvalue of $T$ or a 2-by-2 block with no eigenvalues (see 9.4 and 9.9). Each entry in a 1-by-1 block on the diagonal of $\mathcal{M}(T)$ is an eigenvalue of $T$ and thus makes a contribution to trace $T$. If $\mathcal{M}(T)$ has any 2-by-2 blocks on the diagonal, consider a typical one

$$\begin{bmatrix} a & c \\ b & d \end{bmatrix}.$$

The characteristic polynomial of this 2-by-2 matrix is $(x-a)(x-d)-bc$, which equals

$$x^2 - (a+d)x + (ad - bc).$$

*You should carefully review 9.9 to understand the relationship between eigenpairs and characteristic polynomials of 2-by-2 blocks.*

Thus $(-a - d, ad - bc)$ is an eigenpair of $T$. The negative of the first coordinate of this eigenpair, namely, $a + d$, is the contribution of this block to trace $T$. Note that $a + d$ is the sum of the entries on the diagonal of this block. Thus for any basis of $V$ with respect to which the matrix of $T$ has the block upper-triangular form required by 9.4 and 9.9, trace $T$ equals the sum of the entries on the diagonal.

At this point you should suspect that trace $T$ equals the sum of the diagonal entries of the matrix of $T$ with respect to an arbitrary basis. Remarkably, this turns out to be true. To prove it, let's define the **trace** of a square matrix $A$, denoted trace $A$, to be the sum of the diagonal entries. With this notation, we want to prove that trace $T$ = trace $\mathcal{M}(T, (v_1, \ldots, v_n))$, where $(v_1, \ldots, v_n)$ is an arbitrary basis of $V$. We already know this is true if $(v_1, \ldots, v_n)$ is a basis with respect to which $T$ has an upper-triangular matrix (if $V$ is complex) or an appropriate block upper-triangular matrix (if $V$ is real). We will need the following proposition to prove our trace formula for an arbitrary basis.

**10.9 Proposition:** *If $A$ and $B$ are square matrices of the same size, then*

$$\operatorname{trace}(AB) = \operatorname{trace}(BA).$$

PROOF: Suppose

$$A = \begin{bmatrix} a_{1,1} & \ldots & a_{1,n} \\ \vdots & & \vdots \\ a_{n,1} & \ldots & a_{n,n} \end{bmatrix}, \quad B = \begin{bmatrix} b_{1,1} & \ldots & b_{1,n} \\ \vdots & & \vdots \\ b_{n,1} & \ldots & b_{n,n} \end{bmatrix}.$$

The $j^{\text{th}}$ term on the diagonal of $AB$ equals

$$\sum_{k=1}^{n} a_{j,k} b_{k,j}.$$

Thus

$$\begin{aligned} \operatorname{trace}(AB) &= \sum_{j=1}^{n} \sum_{k=1}^{n} a_{j,k} b_{k,j} \\ &= \sum_{k=1}^{n} \sum_{j=1}^{n} b_{k,j} a_{j,k} \\ &= \sum_{k=1}^{n} k^{\text{th}} \text{ term on the diagonal of } BA \\ &= \operatorname{trace}(BA), \end{aligned}$$

as desired.                                                                    ∎

Now we can prove that the sum of the diagonal entries of the matrix of an operator is independent of the basis with respect to which the matrix is computed.

**10.10 Corollary:** *Suppose $T \in \mathcal{L}(V)$. If $(u_1, \ldots, u_n)$ and $(v_1, \ldots, v_n)$ are bases of $V$, then*

$$\operatorname{trace} \mathcal{M}(T, (u_1, \ldots, u_n)) = \operatorname{trace} \mathcal{M}(T, (v_1, \ldots, v_n)).$$

PROOF:  Suppose $(u_1, \ldots, u_n)$ and $(v_1, \ldots, v_n)$ are bases of $V$. Let $A = \mathcal{M}(I, (u_1, \ldots, u_n), (v_1, \ldots, v_n))$. Then

*The third equality here depends on the associative property of matrix multiplication.*

$$\begin{aligned}
\operatorname{trace} \mathcal{M}(T, (u_1, \ldots, u_n)) &= \operatorname{trace}\left(A^{-1}(\mathcal{M}(T, (v_1, \ldots, v_n))A)\right) \\
&= \operatorname{trace}\left((\mathcal{M}(T, (v_1, \ldots, v_n))A)A^{-1}\right) \\
&= \operatorname{trace} \mathcal{M}(T, (v_1, \ldots, v_n)),
\end{aligned}$$

where the first equality follows from 10.3 and the second equality follows from 10.9. The third equality completes the proof.  ∎

The theorem below states that the trace of an operator equals the sum of the diagonal entries of the matrix of the operator. This theorem does not specify a basis because, by the corollary above, the sum of the diagonal entries of the matrix of an operator is the same for every choice of basis.

**10.11    Theorem:**  *If $T \in \mathcal{L}(V)$, then* $\operatorname{trace} T = \operatorname{trace} \mathcal{M}(T)$.

PROOF:  Let $T \in \mathcal{L}(V)$. As noted above, $\operatorname{trace} \mathcal{M}(T)$ is independent of which basis of $V$ we choose (by 10.10). Thus to show that

$$\operatorname{trace} T = \operatorname{trace} \mathcal{M}(T)$$

for every basis of $V$, we need only show that the equation above holds for some basis of $V$. We already did this (on page 218), choosing a basis of $V$ with respect to which $\mathcal{M}(T)$ is an upper-triangular matrix (if $V$ is a complex vector space) or an appropriate block upper-triangular matrix (if $V$ is a real vector space).  ∎

If we know the matrix of an operator on a complex vector space, the theorem above allows us to find the sum of all the eigenvalues without finding any of the eigenvalues. For example, consider the operator on $\mathbf{C}^5$ whose matrix is

$$\begin{bmatrix}
0 & 0 & 0 & 0 & -3 \\
1 & 0 & 0 & 0 & 6 \\
0 & 1 & 0 & 0 & 0 \\
0 & 0 & 1 & 0 & 0 \\
0 & 0 & 0 & 1 & 0
\end{bmatrix}.$$

No one knows an exact formula for any of the eigenvalues of this operator. However, we do know that the sum of the eigenvalues equals 0 because the sum of the diagonal entries of the matrix above equals 0.

The theorem above also allows us easily to prove some useful properties about traces of operators by shifting to the language of traces of matrices, where certain properties have already been proved or are obvious. We carry out this procedure in the next corollary.

**10.12   Corollary:**  *If $S, T \in \mathcal{L}(V)$, then*

$$\text{trace}(ST) = \text{trace}(TS) \quad and \quad \text{trace}(S + T) = \text{trace} \, S + \text{trace} \, T.$$

PROOF:  Suppose $S, T \in \mathcal{L}(V)$. Choose any basis of $V$. Then

$$\begin{aligned}
\text{trace}(ST) &= \text{trace} \, \mathcal{M}(ST) \\
&= \text{trace}(\mathcal{M}(S)\mathcal{M}(T)) \\
&= \text{trace}(\mathcal{M}(T)\mathcal{M}(S)) \\
&= \text{trace} \, \mathcal{M}(TS) \\
&= \text{trace}(TS),
\end{aligned}$$

where the first and last equalities come from 10.11 and the middle equality comes from 10.9. This completes the proof of the first assertion in the corollary.

To prove the second assertion in the corollary, note that

$$\begin{aligned}
\text{trace}(S + T) &= \text{trace} \, \mathcal{M}(S + T) \\
&= \text{trace}(\mathcal{M}(S) + \mathcal{M}(T)) \\
&= \text{trace} \, \mathcal{M}(S) + \text{trace} \, \mathcal{M}(T) \\
&= \text{trace} \, S + \text{trace} \, T,
\end{aligned}$$

where again the first and last equalities come from 10.11; the third equality is obvious from the definition of the trace of a matrix. This completes the proof of the second assertion in the corollary.            ∎

The techniques we have developed have the following curious corollary. The generalization of this result to infinite-dimensional vector spaces has important consequences in quantum theory.

*The statement of this corollary does not involve traces, though the short proof uses traces. Whenever something like this happens in mathematics, we can be sure that a good definition lurks in the background.*

**10.13   Corollary:**  *There do not exist operators $S, T \in \mathcal{L}(V)$ such that $ST - TS = I$.*

PROOF:  Suppose $S, T \in \mathcal{L}(V)$. Then

$$\text{trace}(ST - TS) = \text{trace}(ST) - \text{trace}(TS)$$
$$= 0,$$

where the second equality comes from 10.12.  Clearly the trace of $I$ equals $\dim V$, which is not 0.  Because $ST - TS$ and $I$ have different traces, they cannot be equal. ∎

## Determinant of an Operator

*Note that $\det T$ depends only on $T$ and not on a basis of $V$ because the characteristic polynomial of $T$ does not depend on a choice of basis.*

For $T \in \mathcal{L}(V)$, we define the ***determinant*** of $T$, denoted $\det T$, to be $(-1)^{\dim V}$ times the constant term in the characteristic polynomial of $T$. The motivation for the factor $(-1)^{\dim V}$ in this definition comes from 10.6.

If $V$ is a complex vector space, then $\det T$ equals the product of the eigenvalues of $T$, counting multiplicity; this follows immediately from 10.6.  Recall that if $V$ is a complex vector space, then there is a basis of $V$ with respect to which $T$ has an upper-triangular matrix (see 5.13); thus $\det T$ equals the product of the diagonal entries of this matrix (see 8.10).

If $V$ is a real vector space, then $\det T$ equals the product of the eigenvalues of $T$ times the product of the second coordinates of the eigenpairs of $T$, each repeated according to multiplicity—this follows from 10.7 and the observation that $m = \dim V - 2M$ (in the notation of 10.7), and hence $(-1)^m = (-1)^{\dim V}$.

For example, suppose $T \in \mathcal{L}(\mathbf{C}^3)$ is the operator whose matrix is given by 10.8. As we noted in the last section, the eigenvalues of $T$ are $1, 2 + 3i$, and $2 - 3i$, each with multiplicity 1. Computing the product of the eigenvalues, we have $\det T = (1)(2 + 3i)(2 - 3i)$; in other words, $\det T = 13$.

As another example, suppose $T \in \mathcal{L}(\mathbf{R}^3)$ is the operator whose matrix is also given by 10.8 (note that in the previous paragraph we were working on a complex vector space; now we are working on a real vector space). Then, as we noted earlier, 1 is the only eigenvalue of $T$ (it has multiplicity 1) and $(-4, 13)$ is the only eigenpair of $T$ (it has multiplicity 1). Computing the product of the eigenvalues times the product of the second coordinates of the eigenpairs, we have $\det T = (1)(13)$; in other words, $\det T = 13$.

The reason that the operators in the two previous examples have the same determinant will become clear after we find a formula (valid on both complex and real vector spaces) for computing the determinant of an operator from its matrix.

In this section, we will prove some simple but important properties of determinants. In the next section, we will discover how to calculate $\det T$ from the matrix of $T$ (with respect to an arbitrary basis). We begin with a crucial result that has an easy proof with our approach.

**10.14**  **Proposition:**  *An operator is invertible if and only if its determinant is nonzero.*

PROOF:  First suppose $V$ is a complex vector space and $T \in \mathcal{L}(V)$. The operator $T$ is invertible if and only if 0 is not an eigenvalue of $T$. Clearly this happens if and only if the product of the eigenvalues of $T$ is not 0. Thus $T$ is invertible if and only if $\det T \neq 0$, as desired.

Now suppose $V$ is a real vector space and $T \in \mathcal{L}(V)$. Again, $T$ is invertible if and only if 0 is not an eigenvalue of $T$. Using the notation of 10.7, we have

**10.15**  $$\det T = \lambda_1 \ldots \lambda_m \beta_1 \ldots \beta_M,$$

where the $\lambda$'s are the eigenvalues of $T$ and the $\beta$'s are the second coordinates of the eigenpairs of $T$, each repeated according to multiplicity. For each eigenpair $(\alpha_j, \beta_j)$, we have $\alpha_j{}^2 < 4\beta_j$. In particular, each $\beta_j$ is positive. This implies (see 10.15) that $\lambda_1 \ldots \lambda_m \neq 0$ if and only if $\det T \neq 0$. Thus $T$ is invertible if and only if $\det T \neq 0$, as desired.  ∎

If $T \in \mathcal{L}(V)$ and $\lambda, z \in \mathbf{F}$, then $\lambda$ is an eigenvalue of $T$ if and only if $z - \lambda$ is an eigenvalue of $zI - T$. This follows from

$$-(T - \lambda I) = (zI - T) - (z - \lambda)I.$$

Raising both sides of this equation to the $\dim V$ power and then taking null spaces of both sides shows that the multiplicity of $\lambda$ as an eigenvalue of $T$ equals the multiplicity of $z - \lambda$ as an eigenvalue of $zI - T$.

The next lemma gives the analogous result for eigenpairs. We will use this lemma to show that the characteristic polynomial can be expressed as a certain determinant.

*Real vector spaces are harder to deal with than complex vector spaces. The first time you read this chapter, you may want to concentrate on the basic ideas by considering only complex vector spaces and ignoring the special procedures needed to deal with real vector spaces.*

**10.16   Lemma:**  *Suppose $V$ is a real vector space, $T \in \mathcal{L}(V)$, and $\alpha, \beta, x \in \mathbf{R}$ with $\alpha^2 < 4\beta$. Then $(\alpha, \beta)$ is an eigenpair of $T$ if and only if $(-2x - \alpha, x^2 + \alpha x + \beta)$ is an eigenpair of $xI - T$. Furthermore, these eigenpairs have the same multiplicities.*

PROOF:  First we need to check that $(-2x - \alpha, x^2 + \alpha x + \beta)$ satisfies the inequality required of an eigenpair. We have

$$(-2x - \alpha)^2 = 4x^2 + 4\alpha x + \alpha^2$$
$$< 4x^2 + 4\alpha x + 4\beta$$
$$= 4(x^2 + \alpha x + \beta).$$

Thus $(-2x - \alpha, x^2 + \alpha x + \beta)$ satisfies the required inequality. Now

$$T^2 + \alpha T + \beta I = (xI - T)^2 - (2x + \alpha)(xI - T) + (x^2 + \alpha x + \beta)I,$$

as you should verify. Thus $(\alpha, \beta)$ is an eigenpair of $T$ if and only if $(-2x - \alpha, x^2 + \alpha x + \beta)$ is an eigenpair of $xI - T$. Furthermore, raising both sides of the equation above to the $\dim V$ power and then taking null spaces of both sides shows that the multiplicities are equal.   ∎

Most textbooks take the theorem below as the definition of the characteristic polynomial. Texts using that approach must spend considerably more time developing the theory of determinants before they get to interesting linear algebra.

**10.17   Theorem:**  *Suppose $T \in \mathcal{L}(V)$. Then the characteristic polynomial of $T$ equals $\det(zI - T)$.*

PROOF:  First suppose $V$ is a complex vector space. Let $\lambda_1, \dots, \lambda_n$ denote the eigenvalues of $T$, repeated according to multiplicity. Thus for $z \in \mathbf{C}$, the eigenvalues of $zI - T$ are $z - \lambda_1, \dots, z - \lambda_n$, repeated according to multiplicity. The determinant of $zI - T$ is the product of these eigenvalues. In other words,

$$\det(zI - T) = (z - \lambda_1) \dots (z - \lambda_n).$$

The right side of the equation above is, by definition, the characteristic polynomial of $T$, completing the proof when $V$ is a complex vector space.

Now suppose $V$ is a real vector space. Let $\lambda_1, \dots, \lambda_m$ denote the eigenvalues of $T$ and let $(\alpha_1, \beta_1), \dots, (\alpha_M, \beta_M)$ denote the eigenpairs of $T$, each repeated according to multiplicity. Thus for $x \in \mathbf{R}$, the eigenvalues of $xI - T$ are $x - \lambda_1, \dots, x - \lambda_m$ and, by 10.16, the eigenpairs of $xI - T$ are

$$(-2x - \alpha_1, x^2 + \alpha_1 x + \beta_1), \dots, (-2x - \alpha_M, x^2 + \alpha_M x + \beta_M),$$

each repeated according to multiplicity. Hence

$$\det(xI - T) = (x - \lambda_1) \dots (x - \lambda_m)(x^2 + \alpha_1 x + \beta_1) \dots (x^2 + \alpha_M x + \beta_M).$$

The right side of the equation above is, by definition, the characteristic polynomial of $T$, completing the proof when $V$ is a real vector space. $\blacksquare$

## Determinant of a Matrix

Most of this section is devoted to discovering how to calculate $\det T$ from the matrix of $T$ (with respect to an arbitrary basis). Let's start with the easiest situation. Suppose $V$ is a complex vector space, $T \in \mathcal{L}(V)$, and we choose a basis of $V$ with respect to which $T$ has an upper-triangular matrix. Then, as we noted in the last section, $\det T$ equals the product of the diagonal entries of this matrix. Could such a simple formula be true in general?

Unfortunately the determinant is more complicated than the trace. In particular, $\det T$ need not equal the product of the diagonal entries of $\mathcal{M}(T)$ with respect to an arbitrary basis. For example, the operator on $\mathbf{F}^3$ whose matrix equals 10.8 has determinant 13, as we saw in the last section. However, the product of the diagonal entries of that matrix equals 0.

For each square matrix $A$, we want to define the determinant of $A$, denoted $\det A$, in such a way that $\det T = \det \mathcal{M}(T)$ regardless of which basis is used to compute $\mathcal{M}(T)$. We begin our search for the correct definition of the determinant of a matrix by calculating the determinants of some special operators.

Let $c_1, \dots, c_n \in \mathbf{F}$ be nonzero scalars and let $(v_1, \dots, v_n)$ be a basis of $V$. Consider the operator $T \in \mathcal{L}(V)$ such that $\mathcal{M}(T, (v_1, \dots, v_n))$ equals

$$10.18 \qquad \begin{bmatrix} 0 & & & & & c_n \\ c_1 & 0 & & & & \\ & c_2 & 0 & & & \\ & & \ddots & \ddots & & \\ & & & c_{n-1} & 0 \end{bmatrix};$$

here all entries of the matrix are 0 except for the upper-right corner and along the line just below the diagonal. Let's find the determinant of $T$. Note that

$$(v_1, Tv_1, T^2 v_1, \ldots, T^{n-1} v_1) = (v_1, c_1 v_2, c_1 c_2 v_3, \ldots, c_1 \ldots c_{n-1} v_n).$$

Thus $(v_1, Tv_1, \ldots, T^{n-1} v_1)$ is linearly independent (the $c$'s are all non-zero). Hence if $p$ is a nonzero polynomial with degree at most $n - 1$, then $p(T)v_1 \neq 0$. In other words, the minimal polynomial of $T$ cannot have degree less than $n$. As you should verify, $T^n v_j = c_1 \ldots c_n v_j$ for each $j$, and hence $T^n = c_1 \ldots c_n I$. Thus $z^n - c_1 \ldots c_n$ is the minimal polynomial of $T$. Because $n = \dim V$, we see that $z^n - c_1 \ldots c_n$ is also the characteristic polynomial of $T$. Multiplying the constant term of this polynomial by $(-1)^n$, we get

*Recall that if the minimal polynomial of an operator $T \in \mathcal{L}(V)$ has degree $\dim V$, then the characteristic polynomial of $T$ equals the minimal polynomial of $T$. Computing the minimal polynomial is often an efficient method of finding the characteristic polynomial.*

$$10.19 \qquad \det T = (-1)^{n-1} c_1 \ldots c_n.$$

If some $c_j$ equals 0, then clearly $T$ is not invertible, so $\det T = 0$ and the same formula holds. Thus in order to have $\det T = \det \mathcal{M}(T)$, we will have to make the determinant of 10.18 equal to $(-1)^{n-1} c_1 \ldots c_n$. However, we do not yet have enough evidence to make a reasonable guess about the proper definition of the determinant of an arbitrary square matrix.

To compute the determinants of a more complicated class of operators, we introduce the notion of permutation. A **permutation** of $(1, \ldots, n)$ is a list $(m_1, \ldots, m_n)$ that contains each of the numbers $1, \ldots, n$ exactly once. The set of all permutations of $(1, \ldots, n)$ is denoted perm $n$. For example, $(2, 3, \ldots, n, 1) \in$ perm $n$. You should think of an element of perm $n$ as a rearrangement of the first $n$ integers.

For simplicity we will work with matrices with complex entries (at this stage we are providing only motivation—formal proofs will come later). Let $c_1, \ldots, c_n \in \mathbf{C}$ and let $(v_1, \ldots, v_n)$ be a basis of $V$, which we are assuming is a complex vector space. Consider a permutation $(p_1, \ldots, p_n) \in$ perm $n$ that can be obtained as follows: break $(1, \ldots, n)$

into lists of consecutive integers and in each list move the first term to the end of that list. For example, taking $n = 9$, the permutation

**10.20**                     $(2, 3, 1, 5, 6, 7, 4, 9, 8)$

is obtained from $(1, 2, 3), (4, 5, 6, 7), (8, 9)$ by moving the first term of each of these lists to the end, producing $(2, 3, 1), (5, 6, 7, 4), (9, 8)$, and then putting these together to form 10.20. Let $T \in \mathcal{L}(V)$ be the operator such that

**10.21**                     $T v_k = c_k v_{p_k}$

for $k = 1, \dots, n$. We want to find a formula for $\det T$. This generalizes our earlier example because if $(p_1, \dots, p_n)$ happens to be the permutation $(2, 3, \dots, n, 1)$, then the operator $T$ whose matrix equals 10.18 is the same as the operator $T$ defined by 10.21.

With respect to the basis $(v_1, \dots, v_n)$, the matrix of the operator $T$ defined by 10.21 is a block diagonal matrix

$$A = \begin{bmatrix} A_1 & & 0 \\ & \ddots & \\ 0 & & A_M \end{bmatrix},$$

where each block is a square matrix of the form 10.18. The eigenvalues of $T$ equal the union of the eigenvalues of $A_1, \dots, A_M$ (see Exercise 3 in Chapter 9). Recalling that the determinant of an operator on a complex vector space is the product of the eigenvalues, we see that our definition of the determinant of a square matrix should force

$$\det A = (\det A_1) \dots (\det A_M).$$

However, we already know how to compute the determinant of each $A_j$, which has the same form as 10.18 (of course with a different value of $n$). Putting all this together, we see that we should have

$$\det A = (-1)^{n_1 - 1} \dots (-1)^{n_M - 1} c_1 \dots c_n,$$

where $A_j$ has size $n_j$-by-$n_j$. The number $(-1)^{n_1 - 1} \dots (-1)^{n_M - 1}$ is called the sign of the permutation $(p_1, \dots, p_n)$, denoted $\mathrm{sign}(p_1, \dots, p_n)$ (this is a temporary definition that we will change to an equivalent definition later, when we define the sign of an arbitrary permutation).

To put this into a form that does not depend on the particular permutation $(p_1, \ldots, p_n)$, let $a_{j,k}$ denote the entry in row $j$, column $k$, of $A$; thus

$$a_{j,k} = \begin{cases} 0 & \text{if } j \neq p_k; \\ c_k & \text{if } j = p_k. \end{cases}$$

Then

**10.22**    $\det A = \displaystyle\sum_{(m_1, \ldots, m_n) \in \text{perm } n} (\text{sign}(m_1, \ldots, m_n)) a_{m_1, 1} \ldots a_{m_n, n},$

because each summand is 0 except the one corresponding to the permutation $(p_1, \ldots, p_n)$.

Consider now an arbitrary matrix $A$ with entry $a_{j,k}$ in row $j$, column $k$. Using the paragraph above as motivation, we guess that $\det A$ should be defined by 10.22. This will turn out to be correct. We can now dispense with the motivation and begin the more formal approach. First we will need to define the sign of an arbitrary permutation.

*Some texts use the unnecessarily fancy term **signum**, which means the same as sign.*

The **sign** of a permutation $(m_1, \ldots, m_n)$ is defined to be 1 if the number of pairs of integers $(j, k)$ with $1 \leq j < k \leq n$ such that $j$ appears after $k$ in the list $(m_1, \ldots, m_n)$ is even and $-1$ if the number of such pairs is odd. In other words, the sign of a permutation equals 1 if the natural order has been changed an even number of times and equals $-1$ if the natural order has been changed an odd number of times. For example, in the permutation $(2, 3, \ldots, n, 1)$ the only pairs $(j, k)$ with $j < k$ that appear with changed order are $(1, 2), (1, 3), \ldots, (1, n)$; because we have $n - 1$ such pairs, the sign of this permutation equals $(-1)^{n-1}$ (note that the same quantity appeared in 10.19).

The permutation $(2, 1, 3, 4)$, which is obtained from the permutation $(1, 2, 3, 4)$ by interchanging the first two entries, has sign $-1$. The next lemma shows that interchanging any two entries of any permutation changes the sign of the permutation.

**10.23  Lemma:** *Interchanging two entries in a permutation multiplies the sign of the permutation by $-1$.*

PROOF: Suppose we have two permutations, where the second permutation is obtained from the first by interchanging two entries. If the two entries that we interchanged were in their natural order in the first permutation, then they no longer are in the second permutation, and vice versa, for a net change (so far) of 1 or $-1$ (both odd numbers) in the number of pairs not in their natural order.

Consider each entry between the two interchanged entries. If an intermediate entry was originally in the natural order with respect to the first interchanged entry, then it no longer is, and vice versa. Similarly, if an intermediate entry was originally in the natural order with respect to the second interchanged entry, then it no longer is, and vice versa. Thus the net change for each intermediate entry in the number of pairs not in their natural order is 2, 0, or $-2$ (all even numbers).

For all the other entries, there is no change in the number of pairs not in their natural order. Thus the total net change in the number of pairs not in their natural order is an odd number. Thus the sign of the second permutation equals $-1$ times the sign of the first permutation. ∎

If $A$ is an $n$-by-$n$ matrix

**10.24**
$$A = \begin{bmatrix} a_{1,1} & \cdots & a_{1,n} \\ \vdots & & \vdots \\ a_{n,1} & \cdots & a_{n,n} \end{bmatrix},$$

then the **determinant** of $A$, denoted $\det A$, is defined by

*Our motivation for this definition comes from 10.22.*

**10.25**    $\det A = \displaystyle\sum_{(m_1,\ldots,m_n)\in\operatorname{perm} n} (\operatorname{sign}(m_1,\ldots,m_n)) a_{m_1,1} \ldots a_{m_n,n}.$

For example, if $A$ is the 1-by-1 matrix $[a_{1,1}]$, then $\det A = a_{1,1}$ because $\operatorname{perm} 1$ has only one element, namely, $(1)$, which has sign 1. For a more interesting example, consider a typical 2-by-2 matrix. Clearly $\operatorname{perm} 2$ has only two elements, namely, $(1,2)$, which has sign 1, and $(2,1)$, which has sign $-1$. Thus

**10.26**    $\det \begin{bmatrix} a_{1,1} & a_{1,2} \\ a_{2,1} & a_{2,2} \end{bmatrix} = a_{1,1}a_{2,2} - a_{2,1}a_{1,2}.$

To make sure you understand this process, you should now find the formula for the determinant of the 3-by-3 matrix

$$\begin{bmatrix} a_{1,1} & a_{1,2} & a_{1,3} \\ a_{2,1} & a_{2,2} & a_{2,3} \\ a_{3,1} & a_{3,2} & a_{3,3} \end{bmatrix}$$

using just the definition given above (do this even if you already know the answer).

*The set $\operatorname{perm} 3$ contains 6 elements. In general, $\operatorname{perm} n$ contains $n!$ elements. Note that $n!$ rapidly grows large as $n$ increases.*

Let's compute the determinant of an upper-triangular matrix

$$A = \begin{bmatrix} a_{1,1} & & * \\ & \ddots & \\ 0 & & a_{n,n} \end{bmatrix}.$$

The permutation $(1, 2, \ldots, n)$ has sign 1 and thus contributes a term of $a_{1,1} \ldots a_{n,n}$ to the sum 10.25 defining $\det A$. Any other permutation $(m_1, \ldots, m_n) \in \operatorname{perm} n$ contains at least one entry $m_j$ with $m_j > j$, which means that $a_{m_j,j} = 0$ (because $A$ is upper triangular). Thus all the other terms in the sum 10.25 defining $\det A$ make no contribution. Hence $\det A = a_{1,1} \ldots a_{n,n}$. In other words, the determinant of an upper-triangular matrix equals the product of the diagonal entries. In particular, this means that if $V$ is a complex vector space, $T \in \mathcal{L}(V)$, and we choose a basis of $V$ with respect to which $\mathcal{M}(T)$ is upper triangular, then $\det T = \det \mathcal{M}(T)$. Our goal is to prove that this holds for every basis of $V$, not just bases that give upper-triangular matrices.

Generalizing the computation from the paragraph above, next we will show that if $A$ is a block upper-triangular matrix

$$A = \begin{bmatrix} A_1 & & * \\ & \ddots & \\ 0 & & A_m \end{bmatrix},$$

where each $A_j$ is a 1-by-1 or 2-by-2 matrix, then

**10.27**              $\det A = (\det A_1) \ldots (\det A_m).$

To prove this, consider an element of $\operatorname{perm} n$. If this permutation moves an index corresponding to a 1-by-1 block on the diagonal any-place else, then the permutation makes no contribution to the sum 10.25 defining $\det A$ (because $A$ is block upper triangular). For a pair of indices corresponding to a 2-by-2 block on the diagonal, the permutation must either leave these indices fixed or interchange them; otherwise again the permutation makes no contribution to the sum 10.25 defining $\det A$ (because $A$ is block upper triangular). These observations, along with the formula 10.26 for the determinant of a 2-by-2 matrix, lead to 10.27. In particular, if $V$ is a real vector space, $T \in \mathcal{L}(V)$, and we choose a basis of $V$ with respect to which $\mathcal{M}(T)$ is a block upper-triangular matrix with 1-by-1 and 2-by-2 blocks on the diagonal as in 9.9, then $\det T = \det \mathcal{M}(T)$.

Our goal is to prove that $\det T = \det \mathcal{M}(T)$ for every $T \in \mathcal{L}(V)$ and every basis of $V$. To do this, we will need to develop some properties of determinants of matrices. The lemma below is the first of the properties we will need.

*An entire book could be devoted just to deriving properties of determinants. Fortunately we need only a few of the basic properties.*

**10.28   Lemma:**   *Suppose $A$ is a square matrix. If $B$ is the matrix obtained from $A$ by interchanging two columns, then*

$$\det A = -\det B.$$

PROOF:   Suppose $A$ is given by 10.24 and $B$ is obtained from $A$ by interchanging two columns. Think of the sum 10.25 defining $\det A$ and the corresponding sum defining $\det B$. The same products of $a$'s appear in both sums, though they correspond to different permutations. The permutation corresponding to a given product of $a$'s when computing $\det B$ is obtained by interchanging two entries in the corresponding permutation when computing $\det A$, thus multiplying the sign of the permutation by $-1$ (see 10.23). Hence $\det A = -\det B$.   ∎

If $T \in \mathcal{L}(V)$ and the matrix of $T$ (with respect to some basis) has two equal columns, then $T$ is not injective and hence $\det T = 0$. Though this comment makes the next lemma plausible, it cannot be used in the proof because we do not yet know that $\det T = \det \mathcal{M}(T)$.

**10.29   Lemma:**   *If $A$ is a square matrix that has two equal columns, then $\det A = 0$.*

PROOF:   Suppose $A$ is a square matrix that has two equal columns. Interchanging the two equal columns of $A$ gives the original matrix $A$. Thus from 10.28 (with $B = A$), we have $\det A = -\det A$, which implies that $\det A = 0$.   ∎

This section is long, so let's pause for a paragraph. The symbols ✳ that appear on the first page of each chapter are decorations intended to take up space so that the first section of the chapter can start on the next page. Chapter 1 has one of these symbols, Chapter 2 has two of them, and so on. The symbols get smaller with each chapter. What you may not have noticed is that the sum of the areas of the symbols at the beginning of each chapter is the same for all chapters. For example, the diameter of each symbol at the beginning of Chapter 10 equals $1/\sqrt{10}$ times the diameter of the symbol in Chapter 1.

We need to introduce notation that will allow us to represent a matrix in terms of its columns. If $A$ is an $n$-by-$n$ matrix

$$A = \begin{bmatrix} a_{1,1} & \cdots & a_{1,n} \\ \vdots & & \vdots \\ a_{n,1} & \cdots & a_{n,n} \end{bmatrix},$$

then we can think of the $k^{\text{th}}$ column of $A$ as an $n$-by-1 matrix

$$a_k = \begin{bmatrix} a_{1,k} \\ \vdots \\ a_{n,k} \end{bmatrix}.$$

We will write $A$ in the form

$$[\ a_1\ \ \cdots\ \ a_n\ ],$$

with the understanding that $a_k$ denotes the $k^{\text{th}}$ column of $A$. With this notation, note that $a_{j,k}$, with two subscripts, denotes an entry of $A$, whereas $a_k$, with one subscript, denotes a column of $A$.

The next lemma shows that a permutation of the columns of a matrix changes the determinant by a factor of the sign of the permutation.

*Some texts define the determinant to be the function defined on the square matrices that is linear as a function of each column separately and that satisfies 10.30 and $\det I = 1$. To prove that such a function exists and that it is unique takes a nontrivial amount of work.*

**10.30  Lemma:** *Suppose $A = [\ a_1\ \ \cdots\ \ a_n\ ]$ is an $n$-by-$n$ matrix. If $(m_1,\ldots,m_n)$ is a permutation, then*

$$\det[\ a_{m_1}\ \ \cdots\ \ a_{m_n}\ ] = (\text{sign}(m_1,\ldots,m_n))\det A.$$

PROOF: Suppose $(m_1,\ldots,m_n) \in \text{perm}\,n$. We can transform the matrix $[\ a_{m_1}\ \ \cdots\ \ a_{m_n}\ ]$ into $A$ through a series of steps. In each step, we interchange two columns and hence multiply the determinant by $-1$ (see 10.28). The number of steps needed equals the number of steps needed to transform the permutation $(m_1,\ldots,m_n)$ into the permutation $(1,\ldots,n)$ by interchanging two entries in each step. The proof is completed by noting that the number of such steps is even if $(m_1,\ldots,m_n)$ has sign 1, odd if $(m_1,\ldots,m_n)$ has sign $-1$ (this follows from 10.23, along with the observation that the permutation $(1,\ldots,n)$ has sign 1). ∎

Let $A = [\ a_1\ \ \cdots\ \ a_n\ ]$. For $1 \le k \le n$, think of all columns of $A$ except the $k^{\text{th}}$ column as fixed. We have

$$\det A = \det[\; a_1 \;\; \ldots \;\; a_k \;\; \ldots \;\; a_n \;],$$

and we can think of $\det A$ as a function of the $k^{\text{th}}$ column $a_k$. This function, which takes $a_k$ to the determinant above, is a linear map from the vector space of $n$-by-$1$ matrices with entries in $\mathbf{F}$ to $\mathbf{F}$. The linearity follows easily from 10.25, where each term in the sum contains precisely one entry from the $k^{\text{th}}$ column of $A$.

Now we are ready to prove one of the key properties about determinants of square matrices. This property will enable us to connect the determinant of an operator with the determinant of its matrix. Note that this proof is considerably more complicated than the proof of the corresponding result about the trace (see 10.9).

**10.31    Theorem:** *If $A$ and $B$ are square matrices of the same size, then*

$$\det(AB) = \det(BA) = (\det A)(\det B).$$

*This theorem was first proved in 1812 by the French mathematicians Jacques Binet and Augustin-Louis Cauchy.*

PROOF: Let $A = [\; a_1 \;\; \ldots \;\; a_n \;]$, where each $a_k$ is an $n$-by-$1$ column of $A$. Let

$$B = \begin{bmatrix} b_{1,1} & \ldots & b_{1,n} \\ \vdots & & \vdots \\ b_{n,1} & \ldots & b_{n,n} \end{bmatrix} = [\; b_1 \;\; \ldots \;\; b_n \;],$$

where each $b_k$ is an $n$-by-$1$ column of $B$. Let $e_k$ denote the $n$-by-$1$ matrix that equals $1$ in the $k^{\text{th}}$ row and $0$ elsewhere. Note that $Ae_k = a_k$ and $Be_k = b_k$. Furthermore, $b_k = \sum_{m=1}^{n} b_{m,k} e_m$.

First we will prove that $\det(AB) = (\det A)(\det B)$. A moment's thought about the definition of matrix multiplication shows that $AB = [\; Ab_1 \;\; \ldots \;\; Ab_n \;]$. Thus

$$\det(AB) = \det[\; Ab_1 \;\; \ldots \;\; Ab_n \;]$$

$$= \det[\; A(\textstyle\sum_{m_1=1}^{n} b_{m_1,1} e_{m_1}) \;\; \ldots \;\; A(\textstyle\sum_{m_n=1}^{n} b_{m_n,n} e_{m_n}) \;]$$

$$= \det[\; \textstyle\sum_{m_1=1}^{n} b_{m_1,1} Ae_{m_1} \;\; \ldots \;\; \textstyle\sum_{m_n=1}^{n} b_{m_n,n} Ae_{m_n} \;]$$

$$= \sum_{m_1=1}^{n} \cdots \sum_{m_n=1}^{n} b_{m_1,1} \ldots b_{m_n,n} \det[\; Ae_{m_1} \;\; \ldots \;\; Ae_{m_n} \;],$$

where the last equality comes from repeated applications of the linearity of det as a function of one column at a time. In the last sum above, all terms in which $m_j = m_k$ for some $j \neq k$ can be ignored because the determinant of a matrix with two equal columns is 0 (by 10.29). Thus instead of summing over all $m_1, \ldots, m_n$ with each $m_j$ taking on values $1, \ldots, n$, we can sum just over the permutations, where the $m_j$'s have distinct values. In other words,

$$
\begin{aligned}
\det(AB) &= \sum_{(m_1,\ldots,m_n) \in \operatorname{perm} n} b_{m_1,1} \ldots b_{m_n,n} \det[\ Ae_{m_1} \ \ldots \ Ae_{m_n}\ ] \\
&= \sum_{(m_1,\ldots,m_n) \in \operatorname{perm} n} b_{m_1,1} \ldots b_{m_n,n} (\operatorname{sign}(m_1,\ldots,m_n)) \det A \\
&= (\det A) \sum_{(m_1,\ldots,m_n) \in \operatorname{perm} n} (\operatorname{sign}(m_1,\ldots,m_n)) b_{m_1,1} \ldots b_{m_n,n} \\
&= (\det A)(\det B),
\end{aligned}
$$

where the second equality comes from 10.30.

In the paragraph above, we proved that $\det(AB) = (\det A)(\det B)$. Interchanging the roles of $A$ and $B$, we have $\det(BA) = (\det B)(\det A)$. The last equation can be rewritten as $\det(BA) = (\det A)(\det B)$, completing the proof.  ∎

Now we can prove that the determinant of the matrix of an operator is independent of the basis with respect to which the matrix is computed.

**10.32  Corollary:**  *Suppose $T \in \mathcal{L}(V)$. If $(u_1, \ldots, u_n)$ and $(v_1, \ldots, v_n)$ are bases of $V$, then*

$$
\det \mathcal{M}(T, (u_1, \ldots, u_n)) = \det \mathcal{M}(T, (v_1, \ldots, v_n)).
$$

*Note the similarity of this proof to the proof of the analogous result about the trace (see 10.10).*

PROOF:  Suppose $(u_1, \ldots, u_n)$ and $(v_1, \ldots, v_n)$ are bases of $V$. Let $A = \mathcal{M}(I, (u_1, \ldots, u_n), (v_1, \ldots, v_n))$. Then

$$
\begin{aligned}
\det \mathcal{M}(T, (u_1, \ldots, u_n)) &= \det\!\left(A^{-1}(\mathcal{M}(T, (v_1, \ldots, v_n))A)\right) \\
&= \det\!\left((\mathcal{M}(T, (v_1, \ldots, v_n))A)A^{-1}\right) \\
&= \det \mathcal{M}(T, (v_1, \ldots, v_n)),
\end{aligned}
$$

where the first equality follows from 10.3 and the second equality follows from 10.31. The third equality completes the proof.  ∎

The theorem below states that the determinant of an operator equals the determinant of the matrix of the operator. This theorem does not specify a basis because, by the corollary above, the determinant of the matrix of an operator is the same for every choice of basis.

**10.33    Theorem:**  *If $T \in \mathcal{L}(V)$, then $\det T = \det \mathcal{M}(T)$.*

PROOF:  Let $T \in \mathcal{L}(V)$. As noted above, 10.32 implies that $\det \mathcal{M}(T)$ is independent of which basis of $V$ we choose. Thus to show that

$$\det T = \det \mathcal{M}(T)$$

for every basis of $V$, we need only show that the equation above holds for some basis of $V$. We already did this (on page 230), choosing a basis of $V$ with respect to which $\mathcal{M}(T)$ is an upper-triangular matrix (if $V$ is a complex vector space) or an appropriate block upper-triangular matrix (if $V$ is a real vector space).                                           ∎

If we know the matrix of an operator on a complex vector space, the theorem above allows us to find the product of all the eigenvalues without finding any of the eigenvalues. For example, consider the operator on $\mathbf{C}^5$ whose matrix is

$$\begin{bmatrix} 0 & 0 & 0 & 0 & -3 \\ 1 & 0 & 0 & 0 & 6 \\ 0 & 1 & 0 & 0 & 0 \\ 0 & 0 & 1 & 0 & 0 \\ 0 & 0 & 0 & 1 & 0 \end{bmatrix}.$$

No one knows an exact formula for any of the eigenvalues of this operator. However, we do know that the product of the eigenvalues equals $-3$ because the determinant of the matrix above equals $-3$.

The theorem above also allows us easily to prove some useful properties about determinants of operators by shifting to the language of determinants of matrices, where certain properties have already been proved or are obvious. We carry out this procedure in the next corollary.

**10.34    Corollary:**  *If $S, T \in \mathcal{L}(V)$, then*

$$\det(ST) = \det(TS) = (\det S)(\det T).$$

PROOF:  Suppose $S, T \in \mathcal{L}(V)$. Choose any basis of $V$. Then

$$\begin{aligned}
\det(ST) &= \det \mathcal{M}(ST) \\
&= \det(\mathcal{M}(S)\mathcal{M}(T)) \\
&= (\det \mathcal{M}(S))(\det \mathcal{M}(T)) \\
&= (\det S)(\det T),
\end{aligned}$$

where the first and last equalities come from 10.33 and the third equality comes from 10.31.

In the paragraph above, we proved that $\det(ST) = (\det S)(\det T)$. Interchanging the roles of $S$ and $T$, we have $\det(TS) = (\det T)(\det S)$. Because multiplication of elements of $\mathbf{F}$ is commutative, the last equation can be rewritten as $\det(TS) = (\det S)(\det T)$, completing the proof.  ∎

## *Volume*

We proved the basic results of linear algebra before introducing determinants in this final chapter. Though determinants have value as a research tool in more advanced subjects, they play little role in basic linear algebra (when the subject is done right). Determinants do have one important application in undergraduate mathematics, namely, in computing certain volumes and integrals. In this final section we will use the linear algebra we have learned to make clear the connection between determinants and these applications. Thus we will be dealing with a part of analysis that uses linear algebra.

*Most applied mathematicians agree that determinants should rarely be used in serious numeric calculations.*

We begin with some purely linear algebra results that will be useful when investigating volumes. Recall that an isometry on an inner-product space is an operator that preserves norms. The next result shows that every isometry has determinant with absolute value 1.

**10.35  Proposition:**  *Suppose that $V$ is an inner-product space. If $S \in \mathcal{L}(V)$ is an isometry, then $|\det S| = 1$.*

PROOF:  Suppose $S \in \mathcal{L}(V)$ is an isometry. First consider the case where $V$ is a complex inner-product space. Then all the eigenvalues of $S$ have absolute value 1 (by 7.37). Thus the product of the eigenvalues of $S$, counting multiplicity, has absolute value one. In other words, $|\det S| = 1$, as desired.

Now suppose $V$ is a real inner-product space. Then there is an ortho-normal basis of $V$ with respect to which $S$ has a block diagonal matrix, where each block on the diagonal is a 1-by-1 matrix containing 1 or $-1$ or a 2-by-2 matrix of the form

**10.36**
$$\begin{bmatrix} \cos\theta & -\sin\theta \\ \sin\theta & \cos\theta \end{bmatrix},$$

with $\theta \in (0, \pi)$ (see 7.38). Note that the constant term of the characteristic polynomial of each matrix of the form 10.36 equals 1 (because $\cos^2\theta + \sin^2\theta = 1$). Hence the second coordinate of every eigenpair of $S$ equals 1. Thus the determinant of $S$ is the product of 1's and $-1$'s. In particular, $|\det S| = 1$, as desired.                                ∎

Suppose $V$ is a real inner-product space and $S \in \mathcal{L}(V)$ is an isometry. By the proposition above, the determinant of $S$ equals 1 or $-1$. Note that
$$\{v \in V : Sv = -v\}$$
is the subspace of $V$ consisting of all eigenvectors of $S$ corresponding to the eigenvalue $-1$ (or is the subspace $\{0\}$ if $-1$ is not an eigenvalue of $S$). Thinking geometrically, we could say that this is the subspace on which $S$ reverses direction. A careful examination of the proof of the last proposition shows that $\det S = 1$ if this subspace has even dimension and $\det S = -1$ if this subspace has odd dimension.

A self-adjoint operator on a real inner-product space has no eigen-pairs (by 7.11). Thus the determinant of a self-adjoint operator on a real inner-product space equals the product of its eigenvalues, count-ing multiplicity (of course, this holds for any operator, self-adjoint or not, on a complex vector space).

Recall that if $V$ is an inner-product space and $T \in \mathcal{L}(V)$, then $T^*T$ is a positive operator and hence has a unique positive square root, de-noted $\sqrt{T^*T}$ (see 7.27 and 7.28). Because $\sqrt{T^*T}$ is positive, all its eigen-values are nonnegative (again, see 7.27), and hence its determinant is nonnegative. Thus in the corollary below, taking the absolute value of $\det \sqrt{T^*T}$ would be superfluous.

**10.37  Corollary:** *Suppose $V$ is an inner-product space. If $T \in \mathcal{L}(V)$, then*
$$|\det T| = \det \sqrt{T^*T}.$$

*Another proof of this
corollary is suggested
in Exercise 24 in this
chapter.*

PROOF:  Suppose $T \in \mathcal{L}(V)$. By the polar decomposition (7.41), there is an isometry $S \in \mathcal{L}(V)$ such that

$$T = S\sqrt{T^*T}.$$

Thus

$$|\det T| = |\det S| \det \sqrt{T^*T}$$
$$= \det \sqrt{T^*T},$$

where the first equality follows from 10.34 and the second equality follows from 10.35.  ∎

Suppose $V$ is a real inner-product space and $T \in \mathcal{L}(V)$ is invertible. The $\det T$ is either positive or negative. A careful examination of the proof of the corollary above can help us attach a geometric meaning to whichever of these possibilities holds. To see this, first apply the real spectral theorem (7.13) to the positive operator $\sqrt{T^*T}$, getting an orthonormal basis $(e_1, \ldots, e_n)$ of $V$ such that $\sqrt{T^*T}e_j = \lambda_j e_j$, where $\lambda_1, \ldots, \lambda_n$ are the eigenvalues of $\sqrt{T^*T}$, repeated according to multiplicity. Because each $\lambda_j$ is positive, $\sqrt{T^*T}$ never reverses direction. Now consider the polar decomposition

$$T = S\sqrt{T^*T},$$

*We are not formally
defining the phrase
"reverses direction"
because these
comments are meant to
be an intuitive aid to
our understanding, not
rigorous mathematics.*

where $S \in \mathcal{L}(V)$ is an isometry. Then $\det T = (\det S)(\det \sqrt{T^*T})$. Thus whether $\det T$ is positive or negative depends on whether $\det S$ is positive or negative. As we saw earlier, this depends on whether the space on which $S$ reverses direction has even or odd dimension. Because $T$ is the product of $S$ and an operator that never reverses direction (namely, $\sqrt{T^*T}$), we can reasonably say that whether $\det T$ is positive or negative depends on whether $T$ reverses vectors an even or an odd number of times.

Now we turn to the question of volume, where we will consider only the real inner-product space $\mathbf{R}^n$ (with its standard inner product). We would like to assign to each subset $\Omega$ of $\mathbf{R}^n$ its $n$-dimensional volume, denoted volume $\Omega$ (when $n = 2$, this is usually called area instead of volume). We begin with cubes, where we have a good intuitive notion of volume. The **cube** in $\mathbf{R}^n$ with side length $r$ and vertex $(x_1, \ldots, x_n) \in \mathbf{R}^n$ is the set

$$\{(y_1,\ldots,y_n) \in \mathbf{R}^n : x_j < y_j < x_j + r \text{ for } j = 1,\ldots,n\};$$

you should verify that when $n = 2$, this gives a square, and that when $n = 3$, it gives a familiar three-dimensional cube. The volume of a cube in $\mathbf{R}^n$ with side length $r$ is defined to be $r^n$. To define the volume of an arbitrary set $\Omega \subset \mathbf{R}^n$, the idea is to write $\Omega$ as a subset of a union of many small cubes, then add up the volumes of these small cubes. As we approximate $\Omega$ more accurately by unions (perhaps infinite unions) of small cubes, we get a better estimate of volume $\Omega$.

*Readers familiar with outer measure will recognize that concept here.*

Rather than take the trouble to make precise this definition of volume, we will work only with an intuitive notion of volume. Our purpose in this book is to understand linear algebra, whereas notions of volume belong to analysis (though as we will soon see, volume is intimately connected with determinants). Thus for the rest of this section we will rely on intuitive notions of volume rather than on a rigorous development, though we shall maintain our usual rigor in the linear algebra parts of what follows. Everything said here about volume will be correct— the intuitive reasons given here can be converted into formally correct proofs using the machinery of analysis.

For $T \in \mathcal{L}(V)$ and $\Omega \subset \mathbf{R}^n$, define $T(\Omega)$ by

$$T(\Omega) = \{Tx : x \in \Omega\}.$$

Our goal is to find a formula for the volume of $T(\Omega)$ in terms of $T$ and the volume of $\Omega$. First let's consider a simple example. Suppose $\lambda_1,\ldots,\lambda_n$ are positive numbers. Define $T \in \mathcal{L}(\mathbf{R}^n)$ by $T(x_1,\ldots,x_n) = (\lambda_1 x_1,\ldots,\lambda_n x_n)$. If $\Omega$ is a cube in $\mathbf{R}^n$ with side length $r$, then $T(\Omega)$ is a box in $\mathbf{R}^n$ with sides of length $\lambda_1 r,\ldots,\lambda_n r$. This box has volume $\lambda_1 \ldots \lambda_n r^n$, whereas the cube $\Omega$ has volume $r^n$. Thus this particular $T$, when applied to a cube, multiplies volumes by a factor of $\lambda_1 \ldots \lambda_n$, which happens to equal $\det T$.

As above, assume that $\lambda_1,\ldots,\lambda_n$ are positive numbers. Now suppose that $(e_1,\ldots,e_n)$ is an orthonormal basis of $\mathbf{R}^n$ and $T$ is the operator on $\mathbf{R}^n$ that satisfies $Te_j = \lambda_j e_j$ for $j = 1,\ldots,n$. In the special case where $(e_1,\ldots,e_n)$ is the standard basis of $\mathbf{R}^n$, this operator is the same one as defined in the paragraph above. Even for an arbitrary orthonormal basis $(e_1,\ldots,e_n)$, this operator has the same behavior as the one in the paragraph above—it multiplies the $j^{\text{th}}$ basis vector by a factor of $\lambda_j$. Thus we can reasonably assume that this operator also multiplies volumes by a factor of $\lambda_1 \ldots \lambda_n$, which again equals $\det T$.

We need one more ingredient before getting to the main result in this section. Suppose $S \in \mathcal{L}(\mathbf{R}^n)$ is an isometry. For $x, y \in \mathbf{R}^n$, we have

$$\|Sx - Sy\| = \|S(x - y)\|$$
$$= \|x - y\|.$$

In other words, $S$ does not change the distance between points. As you can imagine, this means that $S$ does not change volumes. Specifically, if $\Omega \subset \mathbf{R}^n$, then volume $S(\Omega) =$ volume $\Omega$.

Now we can give our pseudoproof that an operator $T \in \mathcal{L}(\mathbf{R}^n)$ changes volumes by a factor of $|\det T|$.

**10.38   Theorem:** *If* $T \in \mathcal{L}(\mathbf{R}^n)$, *then*

$$\text{volume } T(\Omega) = |\det T|(\text{volume } \Omega)$$

*for* $\Omega \subset \mathbf{R}^n$.

PROOF:   First consider the case where $T \in \mathcal{L}(\mathbf{R}^n)$ is a positive operator. Let $\lambda_1, \ldots, \lambda_n$ be the eigenvalues of $T$, repeated according to multiplicity. Each of these eigenvalues is a nonnegative number (see 7.27). By the real spectral theorem (7.13), there is an orthonormal basis $(e_1, \ldots, e_n)$ of $V$ such that $Te_j = \lambda_j e_j$ for each $j$. As discussed above, this implies that $T$ changes volumes by a factor of $\det T$.

Now suppose $T \in \mathcal{L}(\mathbf{R}^n)$ is an arbitrary operator. By the polar decomposition (7.41), there is an isometry $S \in \mathcal{L}(V)$ such that

$$T = S\sqrt{T^*T}.$$

If $\Omega \subset \mathbf{R}^n$, then $T(\Omega) = S\big(\sqrt{T^*T}(\Omega)\big)$. Thus

$$\begin{aligned}
\text{volume } T(\Omega) &= \text{volume } S\big(\sqrt{T^*T}(\Omega)\big) \\
&= \text{volume } \sqrt{T^*T}(\Omega) \\
&= (\det \sqrt{T^*T})(\text{volume } \Omega) \\
&= |\det T|(\text{volume } \Omega),
\end{aligned}$$

where the second equality holds because volumes are not changed by the isometry $S$ (as discussed above), the third equality holds by the previous paragraph (applied to the positive operator $\sqrt{T^*T}$), and the fourth equality holds by 10.37.   ∎

The theorem above leads to the appearance of determinants in the formula for change of variables in multivariable integration. To describe this, we will again be vague and intuitive. If $\Omega \subset \mathbf{R}^n$ and $f$ is a real-valued function (not necessarily linear) on $\Omega$, then the *integral* of $f$ over $\Omega$, denoted $\int_\Omega f$ or $\int_\Omega f(x)\,dx$, is defined by breaking $\Omega$ into pieces small enough so that $f$ is almost constant on each piece. On each piece, multiply the (almost constant) value of $f$ by the volume of the piece, then add up these numbers for all the pieces, getting an approximation to the integral that becomes more accurate as we divide $\Omega$ into finer pieces. Actually $\Omega$ needs to be a reasonable set (for example, open or measurable) and $f$ needs to be a reasonable function (for example, continuous or measurable), but we will not worry about those technicalities. Also, notice that the $x$ in $\int_\Omega f(x)\,dx$ is a dummy variable and could be replaced with any other symbol.

Fix a set $\Omega \subset \mathbf{R}^n$ and a function (not necessarily linear) $\sigma: \Omega \to \mathbf{R}^n$. We will use $\sigma$ to make a change of variables in an integral. Before we can get to that, we need to define the derivative of $\sigma$, a concept that uses linear algebra. For $x \in \Omega$, the *derivative* of $\sigma$ at $x$ is an operator $T \in \mathcal{L}(\mathbf{R}^n)$ such that

$$\lim_{y \to 0} \frac{\|\sigma(x+y) - \sigma(x) - Ty\|}{\|y\|} = 0.$$

*If $n = 1$, then the derivative in this sense is the operator on $\mathbf{R}$ of multiplication by the derivative in the usual sense of one-variable calculus.*

If an operator $T \in \mathcal{L}(\mathbf{R}^n)$ exists satisfying the equation above, then $\sigma$ is said to be *differentiable* at $x$. If $\sigma$ is differentiable at $x$, then there is a unique operator $T \in \mathcal{L}(\mathbf{R}^n)$ satisfying the equation above (we will not prove this). This operator $T$ is denoted $\sigma'(x)$. Intuitively, the idea is that for $x$ fixed and $\|y\|$ small, a good approximation to $\sigma(x+y)$ is $\sigma(x) + (\sigma'(x))(y)$ (note that $\sigma'(x) \in \mathcal{L}(\mathbf{R}^n)$, so this makes sense). Note that for $x$ fixed the addition of the term $\sigma(x)$ does not change volumes. Thus if $\Gamma$ is a small subset of $\Omega$ containing $x$, then volume $\sigma(\Gamma)$ is approximately equal to volume$(\sigma'(x))(\Gamma)$.

Because $\sigma$ is a function from $\Omega$ to $\mathbf{R}^n$, we can write

$$\sigma(x) = (\sigma_1(x), \dots, \sigma_n(x)),$$

where each $\sigma_j$ is a function from $\Omega$ to $\mathbf{R}$. The partial derivative of $\sigma_j$ with respect to the $k^{\text{th}}$ coordinate is denoted $D_k \sigma_j$. Evaluating this partial derivative at a point $x \in \Omega$ gives $D_k \sigma_j(x)$. If $\sigma$ is differentiable at $x$, then the matrix of $\sigma'(x)$ with respect to the standard basis of $\mathbf{R}^n$

LIVERPOOL JOHN MOORES UNIVERSITY
LEARNING SERVICES

contains $D_k \sigma_j(x)$ in row $j$, column $k$ (we will not prove this). In other words,

**10.39**        $$\mathcal{M}(\sigma'(x)) = \begin{bmatrix} D_1 \sigma_1(x) & \dots & D_n \sigma_1(x) \\ \vdots & & \vdots \\ D_1 \sigma_n(x) & \dots & D_n \sigma_n(x) \end{bmatrix}.$$

Suppose that $\sigma$ is differentiable at each point of $\Omega$ and that $\sigma$ is injective on $\Omega$. Let $f$ be a real-valued function defined on $\sigma(\Omega)$. Let $x \in \Omega$ and let $\Gamma$ be a small subset of $\Omega$ containing $x$. As we noted above,

$$\text{volume } \sigma(\Gamma) \approx \text{volume}(\sigma'(x))(\Gamma),$$

where the symbol $\approx$ means "approximately equal to". Using 10.38, this becomes

$$\text{volume } \sigma(\Gamma) \approx |\det \sigma'(x)| (\text{volume } \Gamma).$$

Let $y = \sigma(x)$. Multiply the left side of the equation above by $f(y)$ and the right side by $f(\sigma(x))$ (because $y = \sigma(x)$, these two quantities are equal), getting

**10.40**        $$f(y) \text{ volume } \sigma(\Gamma) \approx f(\sigma(x)) |\det \sigma'(x)| (\text{volume } \Gamma).$$

Now divide $\Omega$ into many small pieces and add the corresponding versions of 10.40, getting

**10.41**        $$\int_{\sigma(\Omega)} f(y)\, dy = \int_\Omega f(\sigma(x)) |\det \sigma'(x)|\, dx.$$

This formula was our goal. It is called a change of variables formula because you can think of $y = \sigma(x)$ as a change of variables.

The key point when making a change of variables is that the factor of $|\det \sigma'(x)|$ must be included, as in the right side of 10.41. We finish up by illustrating this point with two important examples. When $n = 2$, we can use the change of variables induced by polar coordinates. In this case $\sigma$ is defined by

*If you are not familiar with polar and spherical coordinates, skip the remainder of this section.*

$$\sigma(r, \theta) = (r \cos \theta, r \sin \theta),$$

where we have used $r, \theta$ as the coordinates instead of $x_1, x_2$ for reasons that will be obvious to everyone familiar with polar coordinates (and will be a mystery to everyone else). For this choice of $\sigma$, the matrix of partial derivatives corresponding to 10.39 is

$$\begin{bmatrix} \cos\theta & -r\sin\theta \\ \sin\theta & r\cos\theta \end{bmatrix},$$

as you should verify. The determinant of the matrix above equals $r$, thus explaining why a factor of $r$ is needed when computing an integral in polar coordinates.

Finally, when $n = 3$, we can use the change of variables induced by spherical coordinates. In this case $\sigma$ is defined by

$$\sigma(\rho, \varphi, \theta) = (\rho \sin\varphi \cos\theta, \rho \sin\varphi \sin\theta, \rho \cos\varphi),$$

where we have used $\rho, \theta, \varphi$ as the coordinates instead of $x_1, x_2, x_3$ for reasons that will be obvious to everyone familiar with spherical coordinates (and will be a mystery to everyone else). For this choice of $\sigma$, the matrix of partial derivatives corresponding to 10.39 is

$$\begin{bmatrix} \sin\varphi\cos\theta & \rho\cos\varphi\cos\theta & -\rho\sin\varphi\sin\theta \\ \sin\varphi\sin\theta & \rho\cos\varphi\sin\theta & \rho\sin\varphi\cos\theta \\ \cos\varphi & -\rho\sin\varphi & 0 \end{bmatrix},$$

as you should verify. You should also verify that the determinant of the matrix above equals $\rho^2 \sin\varphi$, thus explaining why a factor of $\rho^2 \sin\varphi$ is needed when computing an integral in spherical coordinates.

## *Exercises*

1.  Suppose $T \in \mathcal{L}(V)$ and $(v_1, \ldots, v_n)$ is a basis of $V$. Prove that $\mathcal{M}(T, (v_1, \ldots, v_n))$ is invertible if and only if $T$ is invertible.

2.  Prove that if $A$ and $B$ are square matrices of the same size and $AB = I$, then $BA = I$.

3.  Suppose $T \in \mathcal{L}(V)$ has the same matrix with respect to every basis of $V$. Prove that $T$ is a scalar multiple of the identity operator.

4.  Suppose that $(u_1, \ldots, u_n)$ and $(v_1, \ldots, v_n)$ are bases of $V$. Let $T \in \mathcal{L}(V)$ be the operator such that $Tv_k = u_k$ for $k = 1, \ldots, n$. Prove that

    $$\mathcal{M}(T, (v_1, \ldots, v_n)) = \mathcal{M}(I, (u_1, \ldots, u_n), (v_1, \ldots, v_n)).$$

5.  Prove that if $B$ is a square matrix with complex entries, then there exists an invertible square matrix $A$ with complex entries such that $A^{-1}BA$ is an upper-triangular matrix.

6.  Give an example of a real vector space $V$ and $T \in \mathcal{L}(V)$ such that $\mathrm{trace}(T^2) < 0$.

7.  Suppose $V$ is a real vector space, $T \in \mathcal{L}(V)$, and $V$ has a basis consisting of eigenvectors of $T$. Prove that $\mathrm{trace}(T^2) \geq 0$.

8.  Suppose $V$ is an inner-product space and $v, w \in \mathcal{L}(V)$. Define $T \in \mathcal{L}(V)$ by $Tu = \langle u, v \rangle w$. Find a formula for $\mathrm{trace}\, T$.

9.  Prove that if $P \in \mathcal{L}(V)$ satisfies $P^2 = P$, then $\mathrm{trace}\, P$ is a nonnegative integer.

10.  Prove that if $V$ is an inner-product space and $T \in \mathcal{L}(V)$, then

    $$\mathrm{trace}\, T^* = \overline{\mathrm{trace}\, T}.$$

11.  Suppose $V$ is an inner-product space. Prove that if $T \in \mathcal{L}(V)$ is a positive operator and $\mathrm{trace}\, T = 0$, then $T = 0$.

12. Suppose $T \in \mathcal{L}(\mathbf{C}^3)$ is the operator whose matrix is

$$\begin{bmatrix} 51 & -12 & -21 \\ 60 & -40 & -28 \\ 57 & -68 & 1 \end{bmatrix}.$$

Someone tells you (accurately) that $-48$ and $24$ are eigenvalues of $T$. Without using a computer or writing anything down, find the third eigenvalue of $T$.

13. Prove or give a counterexample: if $T \in \mathcal{L}(V)$ and $c \in \mathbf{F}$, then $\operatorname{trace}(cT) = c \operatorname{trace} T$.

14. Prove or give a counterexample: if $S, T \in \mathcal{L}(V)$, then $\operatorname{trace}(ST) = (\operatorname{trace} S)(\operatorname{trace} T)$.

15. Suppose $T \in \mathcal{L}(V)$. Prove that if $\operatorname{trace}(ST) = 0$ for all $S \in \mathcal{L}(V)$, then $T = 0$.

16. Suppose $V$ is an inner-product space and $T \in \mathcal{L}(V)$. Prove that if $(e_1, \ldots, e_n)$ is an orthonormal basis of $V$, then

$$\operatorname{trace}(T^*T) = \|Te_1\|^2 + \cdots + \|Te_n\|^2.$$

Conclude that the right side of the equation above is independent of which orthonormal basis $(e_1, \ldots, e_n)$ is chosen for $V$.

17. Suppose $V$ is a complex inner-product space and $T \in \mathcal{L}(V)$. Let $\lambda_1, \ldots, \lambda_n$ be the eigenvalues of $T$, repeated according to multiplicity. Suppose

$$\begin{bmatrix} a_{1,1} & \cdots & a_{1,n} \\ \vdots & & \vdots \\ a_{n,1} & \cdots & a_{n,n} \end{bmatrix}$$

is the matrix of $T$ with respect to some orthonormal basis of $V$. Prove that

$$|\lambda_1|^2 + \cdots + |\lambda_n|^2 \le \sum_{k=1}^{n} \sum_{j=1}^{n} |a_{j,k}|^2.$$

18. Suppose $V$ is an inner-product space. Prove that

$$\langle S, T \rangle = \operatorname{trace}(ST^*)$$

defines an inner product on $\mathcal{L}(V)$.

LIVERPOOL
JOHN MOORES UNIVERSITY
AVRIL ROBARTS LRC
TEL. 0151 231 4022

*Exercise 19 fails on infinite-dimensional inner-product spaces, leading to what are called hyponormal operators, which have a well-developed theory.*

19. Suppose $V$ is an inner-product space and $T \in \mathcal{L}(V)$. Prove that if

$$\|T^*v\| \le \|Tv\|$$

for every $v \in V$, then $T$ is normal.

20. Prove or give a counterexample: if $T \in \mathcal{L}(V)$ and $c \in \mathbf{F}$, then $\det(cT) = c^{\dim V} \det T$.

21. Prove or give a counterexample: if $S, T \in \mathcal{L}(V)$, then $\det(S+T) = \det S + \det T$.

22. Suppose $A$ is a block upper-triangular matrix

$$A = \begin{bmatrix} A_1 & & * \\ & \ddots & \\ 0 & & A_m \end{bmatrix},$$

where each $A_j$ along the diagonal is a square matrix. Prove that

$$\det A = (\det A_1) \dots (\det A_m).$$

23. Suppose $A$ is an $n$-by-$n$ matrix with real entries. Let $S \in \mathcal{L}(\mathbf{C}^n)$ denote the operator on $\mathbf{C}^n$ whose matrix equals $A$, and let $T \in \mathcal{L}(\mathbf{R}^n)$ denote the operator on $\mathbf{R}^n$ whose matrix equals $A$. Prove that trace $S$ = trace $T$ and $\det S = \det T$.

24. Suppose $V$ is an inner-product space and $T \in \mathcal{L}(V)$. Prove that

$$\det T^* = \overline{\det T}.$$

Use this to prove that $|\det T| = \det \sqrt{T^*T}$, giving a different proof than was given in 10.37.

25. Let $a, b, c$ be positive numbers. Find the volume of the ellipsoid

$$\{(x, y, z) \in \mathbf{R}^3 : \frac{x^2}{a^2} + \frac{y^2}{b^2} + \frac{z^2}{c^2} < 1\}$$

by finding a set $\Omega \subset \mathbf{R}^3$ whose volume you know and an operator $T \in \mathcal{L}(\mathbf{R}^3)$ such that $T(\Omega)$ equals the ellipsoid above.

# Symbol Index

# Index

absolute value, 69
addition, 9
adjoint, 118

basis, 27
block diagonal matrix, 142
block upper-triangular matrix,
     195

Cauchy-Schwarz inequality,
     104
Cayley-Hamilton theorem for
     complex vector
     spaces, 173
Cayley-Hamilton theorem for
     real vector spaces,
     207
change of basis, 216
characteristic polynomial of a
     2-by-2 matrix, 199
characteristic polynomial of an
     operator on a complex
     vector space, 172
characteristic polynomial of an
     operator on a real
     vector space, 206
characteristic value, 77
closed under addition, 13
closed under scalar
     multiplication, 13
complex conjugate, 69

complex number, 2
complex spectral theorem,
     133
complex vector space, 10
conjugate transpose, 120
coordinate, 4
cube, 238
cube root of an operator,
     159

degree, 22
derivative, 241
determinant of a matrix,
     229
determinant of an operator,
     222
diagonal matrix, 87
diagonal of a matrix, 83
differentiable, 241
dimension, 31
direct sum, 15
divide, 180
division algorithm, 66
dot product, 98

eigenpair, 205
eigenvalue of a matrix, 194
eigenvalue of an operator,
     77
eigenvector, 77

249

# Undergraduate Texts in Mathematics

*(continued from page ii)*

**Lang:** A First Course in Calculus. Fifth edition.

**Lang:** Calculus of Several Variables. Third edition.

**Lang:** Introduction to Linear Algebra. Second edition.

**Lang:** Linear Algebra. Third edition.

**Lang:** Undergraduate Algebra. Second edition.

**Lang:** Undergraduate Analysis.

**Lax/Burstein/Lax:** Calculus with Applications and Computing. Volume 1.

**LeCuyer:** College Mathematics with APL.

**Lidl/Pilz:** Applied Abstract Algebra.

**Macki-Strauss:** Introduction to Optimal Control Theory.

**Malitz:** Introduction to Mathematical Logic.

**Marsden/Weinstein:** Calculus I, II, III. Second edition.

**Martin:** The Foundations of Geometry and the Non-Euclidean Plane.

**Martin:** Transformation Geometry: An Introduction to Symmetry.

**Millman/Parker:** Geometry: A Metric Approach with Models. Second edition.

**Moschovakis:** Notes on Set Theory.

**Owen:** A First Course in the Mathematical Foundations of Thermodynamics.

**Palka:** An Introduction to Complex Function Theory.

**Pedrick:** A First Course in Analysis.

**Peressini/Sullivan/Uhl:** The Mathematics of Nonlinear Programming.

**Prenowitz/Jantosciak:** Join Geometries.

**Priestley:** Calculus: An Historical Approach.

**Protter/Morrey:** A First Course in Real Analysis. Second edition.

**Protter/Morrey:** Intermediate Calculus. Second edition.

**Roman:** An Introduction to Coding and Information Theory.

**Ross:** Elementary Analysis: The Theory of Calculus.

**Samuel:** Projective Geometry. *Readings in Mathematics.*

**Scharlau/Opolka:** From Fermat to Minkowski.

**Sethuraman:** Rings, Fields, and Vector Spaces: An Approach to Geometric Constructability.

**Sigler:** Algebra.

**Silverman/Tate:** Rational Points on Elliptic Curves.

**Simmonds:** A Brief on Tensor Analysis. Second edition.

**Singer/Thorpe:** Lecture Notes on Elementary Topology and Geometry.

**Smith:** Linear Algebra. Second edition.

**Smith:** Primer of Modern Analysis. Second edition.

**Stanton/White:** Constructive Combinatorics.

**Stillwell:** Elements of Algebra: Geometry, Numbers, Equations.

**Stillwell:** Mathematics and Its History.

**Strayer:** Linear Programming and Its Applications.

**Thorpe:** Elementary Topics in Differential Geometry.

**Toth:** Glimpses of Algebra and Geometry.

**Troutman:** Variational Calculus and Optimal Control. Second edition.

**Valenza:** Linear Algebra: An Introduction to Abstract Mathematics.

**Whyburn/Duda:** Dynamic Topology.

**Wilson:** Much Ado About Calculus.

LIVERPOOL
JOHN MOORES UNIVERSITY
AVRIL ROBARTS LRC
TITHEBARN STREET
LIVERPOOL L2 2ER
TEL. 0151 231 4022